DNA TOPOLOGY

Series editors

David Rickwood
Department of Biology, University of Essex, Wivenhoe Park,
Colchester, Essex CO4 3SQ, UK

David Male
Institute of Psychiatry, De Crespigny Park, Denmark Hill,
London SE5 8AF, UK

DNA TOPOLOGY

Andrew D. Bates
Anthony Maxwell

Department of Biochemistry, University of Leicester,
Leicester LE1 7RH

IRL PRESS
—at—
OXFORD UNIVERSITY PRESS

Oxford University Press, Walton Street, Oxford OX2 6DP

Oxford   New York   Toronto
Delhi   Bombay   Calcutta   Madras   Karachi
Kuala Lumpur   Singapore   Hong Kong   Tokyo
Nairobi   Dar es Salaam   Cape Town
Melbourne   Auckland   Madrid
and associated companies in
Berlin   Ibadan

Oxford is a trade mark of Oxford University Press

In Focus is a registered trade mark of the Chancellor, Masters, and Scholars
of the University of Oxford trading as Oxford University Press

Published in the United States
by Oxford University Press Inc., New York

A catalogue record for this book is available from the British Library

Library of Congress Cataloging in Publication Data
Bates, Andrew D.
DNA topology / Andrew D. Bates, Anthony Maxwell.
(In focus)
Includes bibliographical references and index.
1. DNA–Conformation.   I. Maxwell, Anthony.   II. Title.
III. Series: In focus (Oxford, England)
QP624.B38   1993   574.87'3282–dc20   92–27979
ISBN 0–19–963349–5 (pbk.)

Typeset by Footnote Graphics, Warminster, Wiltshire
Printed by Interprint, Malta

Preface

'I only took the regular course.'
'What was that?' inquired Alice.
'Reeling and writhing, of course, to begin with', the Mock Turtle replied;
Lewis Carroll, *Alice's Adventures in Wonderland*

The Mock Turtle's reply, 'reeling and writhing' on the one hand gives a flavour of some of the subject matter of this book and, on the other, describes a common reaction of students when faced with the concepts of DNA topology. Since the description of the DNA double helix in 1953, the importance of DNA and the basic features of its structure have been widely appreciated by both scientists and non-scientists alike. However, it has more recently become clear that there are many deformations of this model structure that have important biological consequences. Prominent amongst these are the 'topological' deformations: supercoiling, knotting, and catenation. Unfortunately, a full appreciation of these aspects of DNA structure requires the grasp of concepts that can prove difficult for both students and more advanced researchers. We have written this book with the aim of explaining these ideas simply to allow a wider appreciation of DNA topology. We begin with a basic account of DNA structure before going on to cover DNA supercoiling, the definitions of linking number, twist and writhe and the free energy associated with supercoiling. The rather more complex description of DNA lying on a curved surface and its application to the nucleosome is then considered, followed by the phenomena of knotting and catenation. The final chapters deal with the enzymes that alter DNA topology (DNA topoisomerases) and, most importantly, the biological significance of the topological aspects of DNA structure.

We are grateful to Steve Halford for his careful reading of the manuscript and for many helpful comments and suggestions. We acknowledge Geoff Turnock, Dave Weiner, and Chris Willmott for their comments and Mike Dampier for useful discussions. We also thank the authors and publishers who have allowed the reproduction of figures from previous publications. A.D.B. is a Wellcome Trust Postdoctoral Research Fellow; A.M. is a Lister Institute Jenner Fellow.

Andy Bates
Tony Maxwell

Contents

3. DNA on surfaces

4. Knots and catenanes

5. DNA topoisomerases

6. Biological consequences of DNA topology

Abbreviations

A	adenine
Å	Ångstroms
ATP	adenosine triphosphate
att	attachment site
C	cytosine
Ca	catenane number
CAP	catabolite activator protein
ΔLk	linking difference
DNA	deoxyribonucleic acid
DNase I	deoxyribonuclease I
EtBr	ethidium bromide
Φ	DNA winding number
G	free energy
G	guanine
γ	superhelix winding angle
h	helical repeat
h_s	surface-related helical repeat
h_t	twist-related helical repeat
h°	helical repeat under standard conditions
Int	bacteriophage lambda integration protein
K	elastic constant
K'	equilibrium constant
kDa	kilodaltons
Kn	knot number
λ	bacteriophage lambda
Lk	linking number
Lk_m	Lk of most abundant relaxed topoisomer
Lk°	average linking number of relaxed DNA
mRNA	messenger RNA
N	length of DNA in base pairs
PAGE	polyacrylamide gel electrophoresis
R	gas constant
res	site of recombination by resolvase
RNA	ribonucleic acid

Abbreviations

σ	specific linking difference
SLk	surface linking number
STw	surface twist
T	absolute temperature
T	thymine
Tw	twist
ω	angular displacement
Wr	writhe

1

DNA structure

1. Introduction

Of all biological molecules, DNA has perhaps most fired the imagination of scientists and non-scientists alike. This phenomenon probably derives from the aesthetic appeal of the DNA double helix, a structure that is on the one hand apparently simple and, on the other hand, profound in terms of its implications for biological function. The term 'DNA structure' implies the chemical and stereo-chemical details of DNA molecules. The principal purpose of this book is to focus on the higher order structural features of DNA, namely supercoiling, knotting, and catenation. However, before embarking on these topics, it is necessary to consider the primary and secondary structural features, that is to say, how DNA chains are put together and how they can form helical structures.

2. DNA structures

2.1. The Watson–Crick model

Although it is an over-used cliché, it is nevertheless true to say that the elucidation of the structure of the DNA double helix by Watson and Crick in 1953 represents one of the most important scientific discoveries of the twentieth century (1). The impact of their model has been enormous and it constitutes the foundation stone of modern molecular biology. Not only did this structure satisfy the known physical and chemical properties of DNA, it also implied how DNA can fulfil its biological functions. For example, the complementarity of the two strands of the DNA double helix provided a potential mechanism for it to be copied. Further, the defined sequence of bases along a DNA strand enables the encoding of protein sequence information.

DNA is made up of repeated units, nucleotides, comprising three components: a sugar (2'-deoxyribose), phosphate, and one of four heterocyclic bases. There are two purine bases, adenine and guanine, and two pyrimidines, thymine and cytosine. The structures of these components are shown in *Figure 1.1*. The

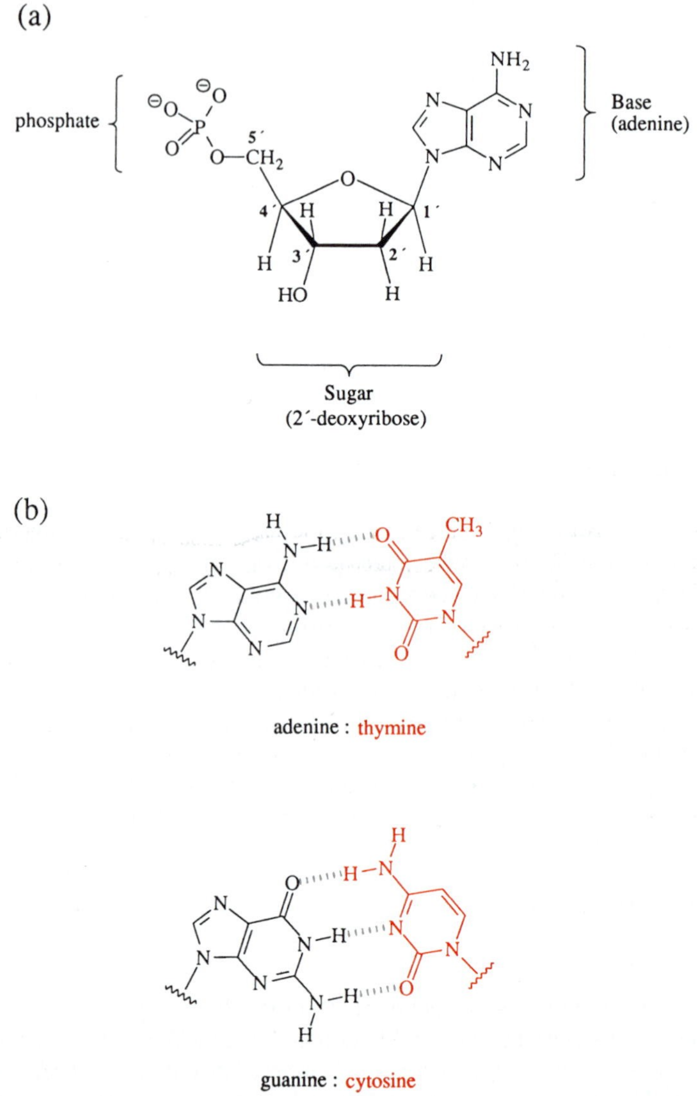

(a)

(b)

adenine : thymine

guanine : cytosine

Figure 1.1. DNA components. **(a)** A nucleotide, the repeat unit of DNA. **(b)** Base pairing in DNA. Purine bases (adenine and guanine) are shown in black, pyrimidine bases (thymine and cytosine) are shown in orange.

bases are attached to the 1′ position of the sugar and a DNA chain is made by joining the sugars via a 3′–5′ phosphodiester linkage, which connects the 3′-hydroxyl group of one sugar to the 5′-hydroxyl group of the next. This constitutes a polynucleotide chain and represents the primary structure of DNA (*Figure 1.2a*).

The principal experimental method used to elucidate the secondary structure of DNA was X-ray diffraction of DNA fibres. This involves taking a highly concentrated solution of DNA and stretching it into a fibre which results in the orientation of the molecules. The diffraction pattern of the mounted fibre is then recorded in an atmosphere of controlled humidity. Such experiments can only provide limited information, but, coupled with chemical data and model building, they led to the discovery of the DNA double helix. The essence of double-stranded DNA is the pairing of the bases by hydrogen bonding (*Figure 1.1b*). Adenine pairs with thymine via two hydrogen bonds and guanine pairs with cytosine via three hydrogen bonds. The two strands in duplex DNA are anti-parallel; that is the polarity of the phosphodiester bonds is $5' \rightarrow 3'$ in one strand and $3' \rightarrow 5'$ in the other (*Figure 1.2b*). In the DNA double helix the two anti-parallel polynucleotide strands are coiled around each other in a right-handed fashion. (A right-handed helix observed along its helix axis spirals in a clockwise fashion away from the observer.) This produces a structure with a largely hydrophobic interior comprising the DNA bases, and a hydrophilic exterior comprising the sugar–phosphate backbone. In the Watson–Crick model of DNA there are 10 base pairs for every turn of the helix and a repeat distance between each successive turn of 3.4 nm (34 Å). The width of the double helix is approximately 2 nm (20 Å) and the bases are perpendicular to the helix axis (*Figure 1.2b*). This form of the DNA double helix is known as the B-form, as distinct from the A-form which was also detected by fibre diffraction experiments but under conditions of lower humidity (see Sections 2.2 and 2.3).

The DNA double helix is a relatively stable structure; however conditions such as high temperature or extremes of pH can cause disruption of the double helix into its single-strand components, a process known as denaturation. When this occurs the hydrogen bond donor and acceptor groups of the DNA bases become exposed. In the course of its biological functions, transient denaturation of DNA occurs and this process can be facilitated by proteins and, as described in Chapter 2, by DNA supercoiling.

The basic description of B-form DNA derived from fibre diffraction studies suffices as a working model for many purposes. However, it is now known that the concept of DNA as a uniform structure is an oversimplification. A number of biochemical and biophysical techniques, in particular X-ray diffraction studies of crystalline DNA, have shown that double-stranded DNA can adopt a variety of conformations. These include local variations in the helical parameters of B-form DNA, other helical species of DNA, and alternatives to the double-helical structure.

2.2. B-form DNA

B-form DNA is thought to represent the conformation of most DNA found in cells. The basic structural parameters were derived from X-ray diffraction data of DNA fibres at high humidity (e.g. 92%). The features which distinguish B-DNA from other forms are the location of the base pairs on the helix axis, the near-perpendicular orientation of the base pairs relative to the helix axis, and distinct major and minor grooves (the former in particular allowing easy access to

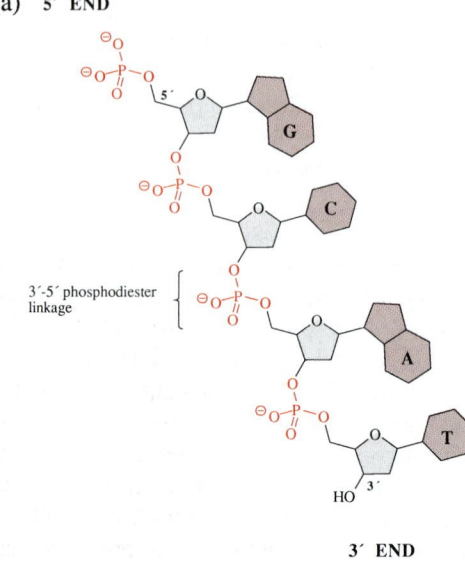

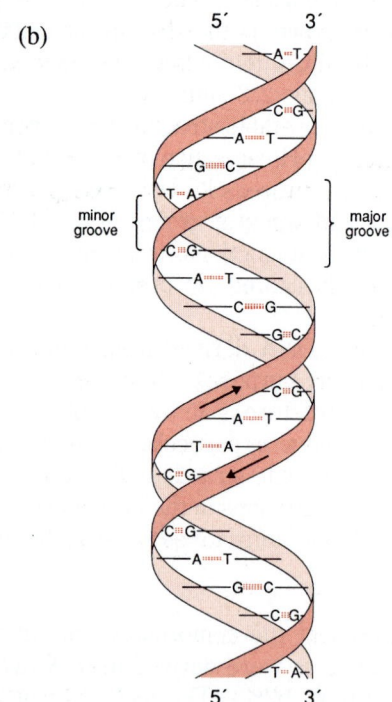

Figure 1.2. DNA structure. **(a)** A polynucleotide chain, showing the phosphodiester bonds which connect adjacent nucleotide units. **(b)** A schematic representation of the B-DNA double helix.

the bases) (*Figure 1.3*). Some of the helical parameters of B-form and other forms of DNA are given in *Table 1.1*. The helical repeat (bp/turn) of mixed-sequence B-form DNA has been determined by a variety of experimental methods and has been found to be close to 10.5 bp/turn; this value depends on the solution conditions, but is often taken as an average for B-DNA (see Chapter 2, Section 3.1). A number of other parameters are required for a complete description of the conformation of the double helix; see Dickerson (2) for a comprehensive description.

In 1980 the structure of a B-form DNA molecule was analysed to atomic resolution using X-ray diffraction of single crystals (3). The molecule used in these studies was the dodecamer d(CGCGAATTCGCG), which is self-complementary (i.e. two molecules can base pair with each other by hydrogen bonding) and forms a double helix of the type suggested by Watson and Crick. Although the overall features of the dodecamer structure conform closely to those expected of B-form DNA, the molecule shows local sequence-dependent variations. For example, the distance between base pairs varies from 0.314 nm to 0.356 nm, the average being about 0.33 nm (3.3 Å). The DNA base pairs are not all perpendicular to the helix axis and indeed several show 'propeller twist' where the purine and pyrimidine pair do not lie flat but are twisted with respect to each other like the blades of a propeller. In addition, the helix axis itself is not straight but is curved. (As discussed in Section 4.1, DNA curvature is an important feature of DNA structure particularly from the stand-point of its biological functions.)

As well as the local conformational variations in B-DNA which have emerged from the studies of the dodecamer and similar short DNA molecules of defined sequence, other workers have described alternative B-like conformations which are given distinct designations such as B', C, D, and E; these species will not be discussed here (4).

2.3. A-form DNA

A-form DNA was identified from X-ray fibre diffraction studies at low humidity (75%) (5). It is not yet clear whether this form of DNA occurs *in vivo*, although

Table 1.1. Structural features of DNA

DNA conformation	Helix handedness	Bp/ turn	Helix diameter (nm)	Major groove	Minor groove
B	Right	10.5	~2.0	Wide and deep	Narrow and deep
A	Right	11.0	~2.6	Narrow and deep	Wide and shallow
Z	Left	12.0	~1.8	Flat	Narrow and deep

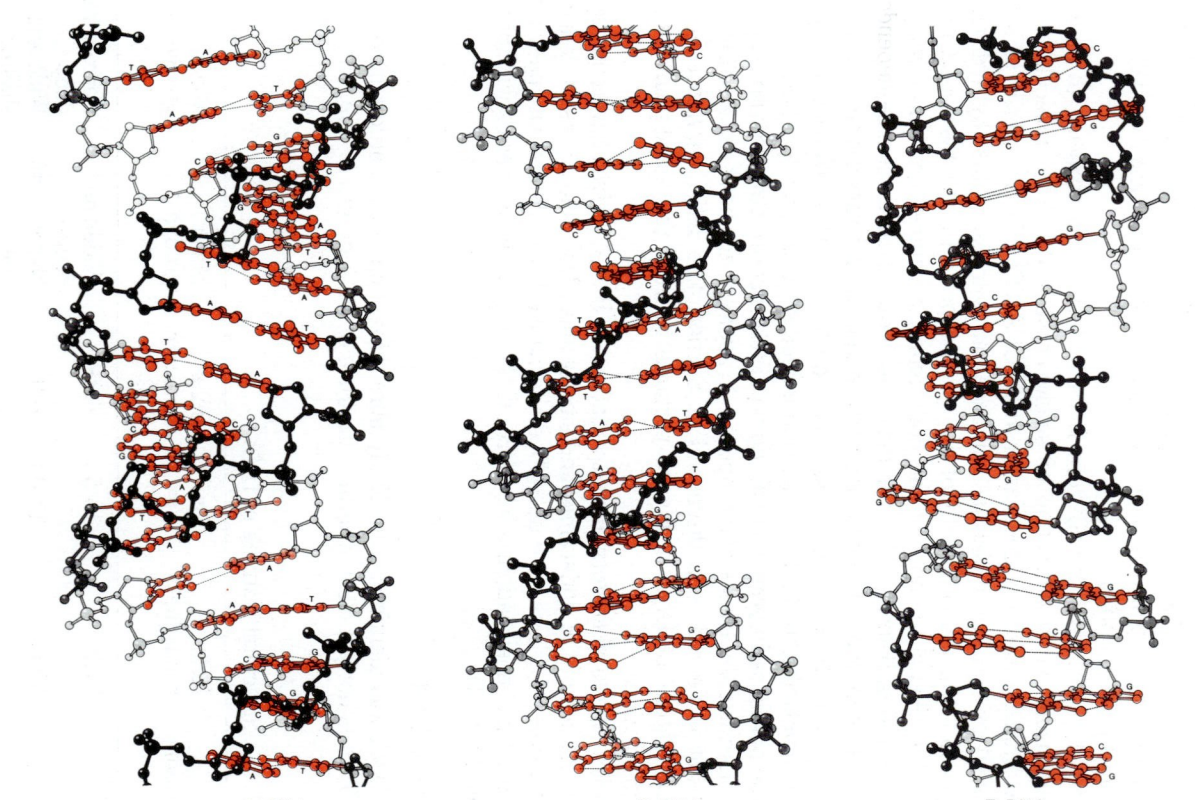

Figure 1.3. Comparison of the A, B, and Z forms of DNA (from (32), Armstrong (1989) *Biochemistry*, 3rd edn. Copyright © 1979, 1983, 1989 by Oxford University Press, Inc.).

the presence of A-DNA has been proposed at specific sites, such as promoter regions (6) and transcription factor binding sites (7). Compared with B-DNA, the A-form helix is broader (2.6 nm in diameter) and the bases are tilted and lie well off the helix axis. The minor groove is thin and deep while the major groove is shallow (*Figure 1.3*). The essential distinction between the A- and B-forms is the conformation of the ribose sugar (the ring 'pucker'), which is $C_{3'}$-*endo* for A-DNA and $C_{2'}$-*endo* for B-DNA. The conformation of A-DNA closely resembles that of a double-helical form of RNA, A-RNA. Several short DNA molecules have been crystallized and shown to conform closely to the A-DNA parameters derived from fibre studies, these include d(CCGG) and d(GGTATACC) (6, 8). Sequence-dependent perturbations in A-form DNA appear to be less marked than in the B-form.

2.4. Z-form DNA

The first DNA molecules to be analysed by X-ray crystallography were the self-complementary oligonucleotides d(CGCGCG) and d(CGCG) in 1979 and 1980 (9, 10). The resulting structures came as a great surprise as the double helices formed were left-handed. This conformation has become known as Z-DNA and it transpired that evidence for its occurrence had been in existence since 1972 from spectroscopic studies of poly(dG–dC) (11). Apart from its difference in handedness and in helical parameters compared to A- and B-forms (*Table 1.1*), Z-form DNA is also distinguished by the zigzag path of the sugar–phosphate backbone (*Figure 1.3*). The Z conformation can be found in DNA molecules with alternating purine–pyrimidine sequences and is favoured by conditions such as high salt. An important issue is whether Z-DNA occurs in nature. At present this remains a controversial question to which we have no definitive answer. One possibility is that left-handed DNA could arise transiently during the course of cellular processes (see Chapter 6, Section 3.1). In the context of DNA topology, the potential of sequences to form Z-DNA can have important consequences. For example, it can be shown that short segments of poly(dG–dC) in plasmids can be induced to form left-handed DNA at sufficient levels of negative supercoiling. This is one example of how the free energy of supercoiling can be used to stabilize DNA conformations that would be unfavourable in the absence of superhelical stress (see Chapter 2, Section 7 and Chapter 6, Section 1.1).

3. Alternative DNA structures

Aside from the ability of DNA to adopt different helical forms, it can also be found as structures which depart from the familiar double helix. Some of these structures are stabilized by DNA supercoiling (e.g. cruciforms and H-DNA). Other structures, such as the anti-parallel quadruple helical structures proposed for chromosome telomeres (12), and the four-stranded structures proposed as recombination intermediates (13), will not be discussed here.

3.1. Cruciforms and Holliday junctions

Cruciforms arise as a consequence of intra-strand base pairing in DNA and consist of a pair of stem and loop or 'hairpin' structures. The formation requires the presence of inverted repeats (also termed 'palindromes') in double-stranded DNA, where a sequence is followed immediately, or soon after, by the same sequence in the opposite orientation. An example of such a sequence occurs in the bacterial plasmid pBR322 (*Figure 1.4*) and consists of an 11 bp inverted repeat spaced by three base pairs (14). This sequence can, in principle, form a cruciform with 11 intra-strand base pairs in the stems. Important questions concern the existence in nature and the biological significance of cruciform structures. It is clear that the formation of a cruciform structure in linear B-DNA will be thermodynamically unfavourable due to the presence of unpaired bases. However, it is possible to show experimentally that cruciforms may form in negatively supercoiled closed-circular DNA *in vitro*. Here the unfavourable free energy associated with negative supercoiling is dissipated by the formation of the cruciform (see Chapter 6, Section 1.1).

Evidence for the presence of cruciforms in such experiments relies on the preference of single-strand-specific reagents (e.g. nuclease S1, bromoacetaldehyde) to act at the site of the inverted repeat sequence. Using these reagents it can be shown that increasing superhelical density promotes the extrusion (looping out) of the cruciform. Whether cruciforms actually exist *in vivo* is a contentious issue. One possibility is that intracellular superhelical densities may be too low to extrude inverted repeat sequences efficiently (15). Kinetic studies of certain cruciforms have suggested that the extrusion process may be too slow for the cruciforms to be of physiological significance (16, 17).

A Holliday junction is a four-way junction formed between two DNA duplexes during homologous recombination (*Figure 1.5*). In this process, two homologous DNA double helices are aligned and, as a result of cleavage of one strand of each helix, base-pairing is established with the intact strand of the other helix (18). Resealing of the DNA strands results in the formation of the Holliday junction which may be redrawn so that it resembles a cruciform without the two unpaired DNA loops (*Figure 1.5c*). During recombination, subsequent cleavage of the other strand of each helix and resealing of the broken ends will generate recombinant DNA molecules.

One way in which Holliday junctions have been studied is by the synthesis of four separate oligonucleotides which, when base-paired, form an artificial Holliday junction (19). In the presence of Mg^{2+} these synthetic junctions have been shown to adopt a 'scissor-like' conformation stabilized by base stacking across the four-way junction (*Figure 1.5d*). The identity of the bases at the four-way junction determines which arms of the structure are co-linear.

3.2. H-DNA

Another potential alternative to the double-stranded helix is 'H-DNA', so-called because of its requirement for protons, but it can also be thought of as 'hinged'

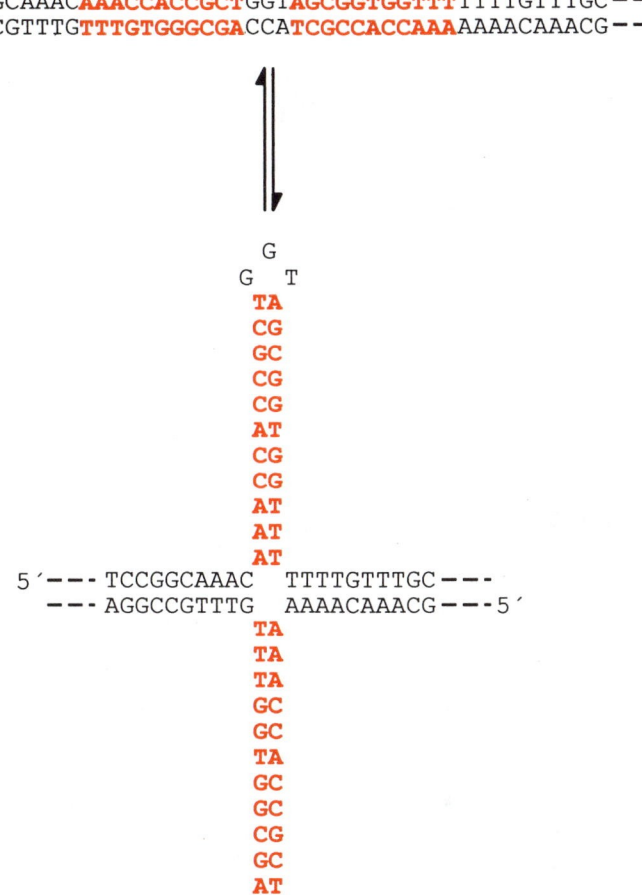

Figure 1.4. Formation of a cruciform from an 'inverted repeat' sequence in DNA. The sequence shown occurs in the bacterial plasmid pBR322.

DNA. This structure was first detected by the unusual sensitivity of certain plasmids to nuclease S1 (20). It is found in the repeating DNA co-polymer $(dT\text{--}dC)_n\cdot(dA\text{-}dG)_n$, which has been found to occur in eukaryotic DNA (20, 21). Although the exact structure is not known, it is thought to consist of a triple-stranded and a single-stranded region of DNA which together form a kink, as shown in *Figure 1.6* (22). The essence of this structure is that a polypyrimidine strand dissociates from the Watson–Crick duplex and lies in the major groove of another section of duplex making non-Watson–Crick base pairs with the purine bases. The original polypurine partner to the polypyrimidine strand remains single-stranded. This structural transition is favoured by negative supercoiling (due to the consequent loss of twist, see Chapter 6, Section 1.1) and low pH (which favours the protonation required for alternative base pairs).

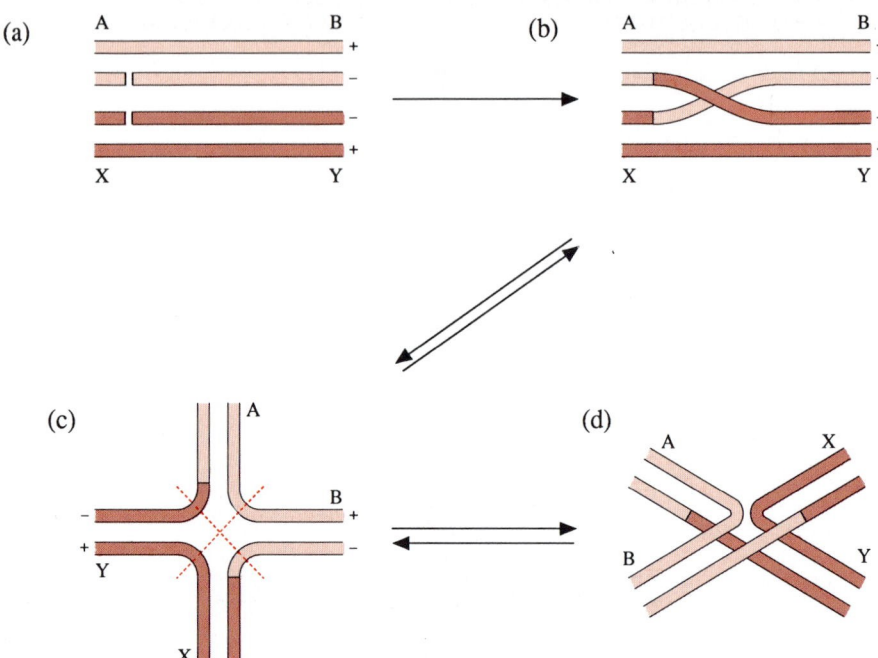

Figure 1.5. Formation of a Holliday junction during homologous recombination. Two double-stranded molecules (shown in different shades) are aligned and, after cleavage of one strand of each duplex: (**a**), a process of strand exchange takes place: (**b**). This intermediate may be redrawn so that it resembles a cruciform: (**c**). Resolution of this intermediate can occur by cleavage and strand exchange across either diagonal in (c). Experiments have shown that, in the presence of magnesium ions, this intermediate probably adopts a scissor-like conformation (19): (**d**).

4. Intrinsic curvature and DNA bending

It is apparent from the above sections that DNA can exhibit a high degree of conformational variability. One of the most striking manifestations of this is the curvature of the helix axis. This is observed in two similar but distinct phenomena, intrinsic curvature of DNA and DNA bending or flexibility. Intrinsic curvature refers to the deformation of the DNA helix axis arising from the preferred conformation of particular DNA sequences. DNA flexibility refers to the ease with which certain DNA sequences can be bent, for example by being wrapped around a protein. A flexible DNA sequence will not necessarily possess any intrinsic curvature in free solution.

4.1. Intrinsic curvature

Intrinsic curvature of DNA can be demonstrated by polyacrylamide gel electrophoresis (PAGE). Curved DNA molecules show reduced electrophoretic mobilities when compared with straight molecules of the same size. Naively this can

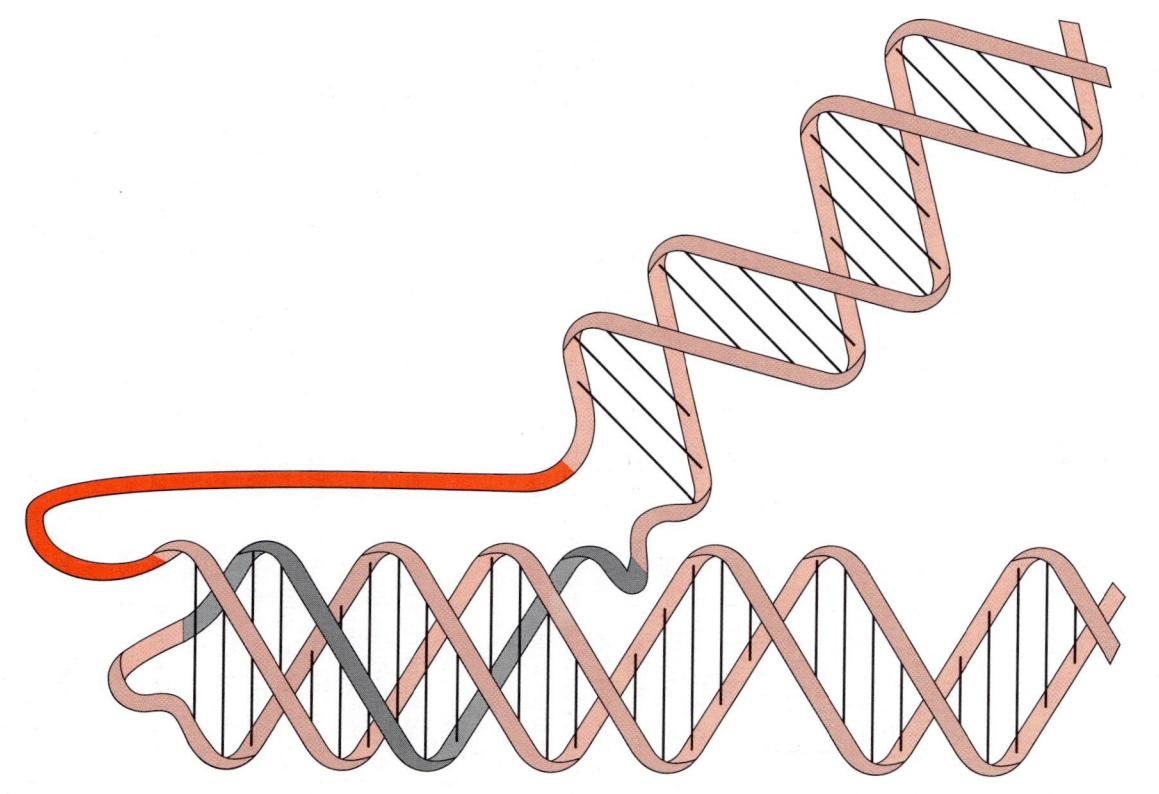

Figure 1.6. A possible structure for H-DNA. The polypyrimidine strand (lighter colour) lies in the major groove of the section of double helix on the left. Its original partner, a polypurine strand (darker colour) remains single-stranded. The whole structure has the form of a kink (redrawn from Htun and Dahlberg (22). Copyright 1989 by the AAAS).

be correlated with the sieving effect of polyacrylamide gels, whereby curved molecules will pass through the pores less easily than their straight counterparts. Intrinsic curvature can arise in DNA sequences which have short runs of 'A's periodically repeated at intervals of 10–11 bp (i.e. the helical repeat of B-form DNA). These runs of As show a structure distinct from random sequence DNA and, when phased, contribute to stable DNA curvature. Such sequences occur naturally, for example in the kinetoplast DNA of trypanosomes where the DNA molecules form small DNA circles (down to 700 bp in size) (23). Indeed it is possible to synthesize artificial DNA sequences with phased A-tracts and make DNA circles in the test-tube as small as 105 bp (24).

A convincing demonstration of the effect of phased A-tracts in DNA curvature was given by the following experiment (25). A series of 423 bp DNA fragments was prepared from the kinetoplast DNA of trypanosomes such that they were circularly permuted, that is, they all had the same sequence but started and finished at different positions (e.g. ABCDE, BCDEA, CDEAB, etc.). Following PAGE it was found that this series of molecules, although identical in molecular weight, differed significantly in their electrophoretic mobilities. These differences could be correlated with the position of a region containing phased A-tracts (*Figure 1.7*); when this region was close to the centre of a DNA fragment the electrophoretic mobility was the lowest, whereas when the A-tract region was close to the end the mobility was the greatest (*Figure 1.7*). Such experiments identify the phased A-tracts as a source of intrinsic curvature in DNA. Other experiments have probed the spacing of the A-tracts and concluded that curvature is maximal when the spacing is close to the helical repeat of B-DNA (26).

4.2. DNA flexibility

DNA flexibility can be of two types, isotropic and anisotropic. Isotropic flexibility means that the DNA molecule can bend equally in all directions whereas anisotropic flexibility means that the DNA has 'hinges' at which it can bend in a preferred direction. Flexibility is a function of the DNA sequence and there can be large differences in the relative ease with which different DNA sequences can be bent. DNA flexibility is relevant to DNA–protein interactions as many DNA-binding proteins bend DNA. For example, the bacteriophage 434 repressor protein has been shown to bend its 14 bp operator DNA sequence (27). The central four bases of the operator sequence, d(ATAT), are not directly in contact with the protein. If these bases are replaced by the sequence d(AGCT) the repressor binding affinity is weakened by a factor of 50. Replacement by d(AAAA) increases repressor affinity 5-fold. These changes in affinity may be correlated with the ease of flexure of the central four base pairs, although other factors, such as an alteration in DNA twist, may also be involved (see Chapter 6, Section 4.2).

Perhaps a more striking example of the influence of DNA flexibility on DNA–protein interactions comes from studies of the *Escherichia coli* catabolite activator protein (CAP), also known as CRP (cAMP receptor protein). CAP activates the

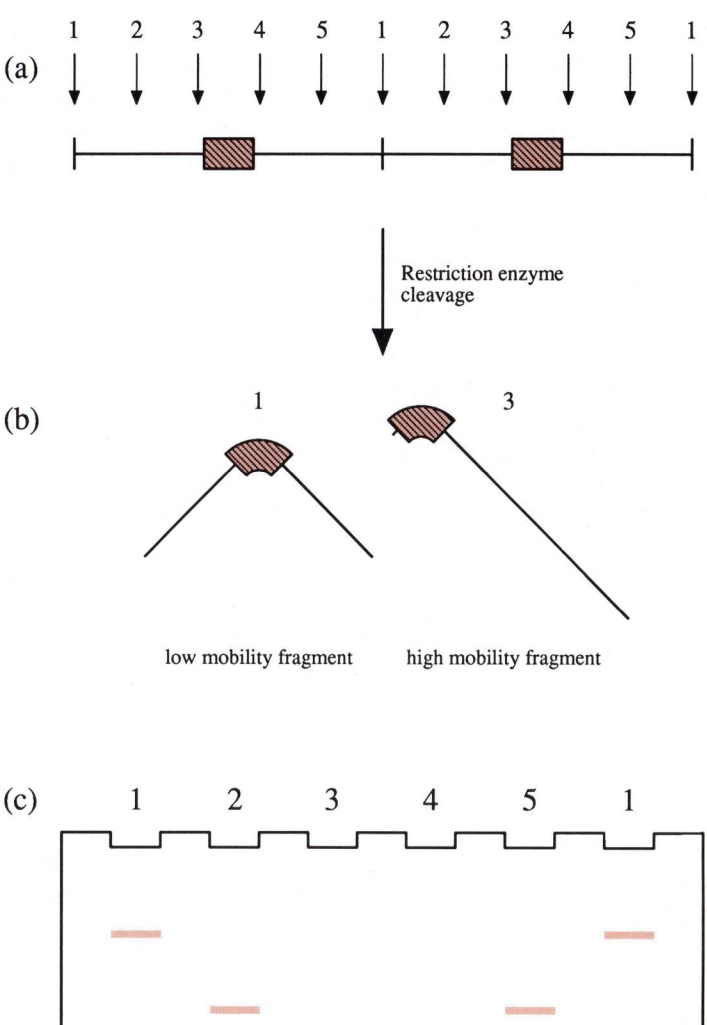

Figure 1.7. Demonstration of the effect of A-tracts in DNA curvature. DNA molecules containing phased A-tract sequences (orange boxes) are cut with a series of restriction enzymes (1–5 in **(a)**). When analysed by polyacrylamide gel electrophoresis **(c)**, the fragments show varying mobility depending on the position of the A-tract sequences **(b)**.

transcription of a number of genes in response to carbon source limitation, which elevates cellular cAMP levels. This results in the formation of the CAP–cAMP complex, which binds to specific DNA sequences at the target promoters (28). In certain cases, CAP can also act as a repressor. A variety of biochemical data have suggested that the interaction of CAP with DNA induces bending of the DNA helix. This has now been confirmed by the X-ray-derived structure of the CAP–DNA complex, which reveals a bend of approximately 90° in the DNA (29). Gartenburg and Crothers have shown that changes in DNA sequence outside the consensus CAP binding site can change the binding affinity 10-fold, and alter the bending angle by up to 30° (30). These effects can be attributed to the influence of certain DNA sequences on the ease with which the DNA can be bent.

It is thought that the bending of the DNA is involved in transcriptional activation by CAP, perhaps by facilitating the interaction of upstream sequences with RNA polymerase. Such an idea is supported by experiments showing that CAP binding sites can be replaced by A-tract sequences, which can impart intrinsic curvature to DNA (see Section 4.1), resulting in transcriptional activation in a phase-dependent manner (31).

It is clear that DNA flexibility will have a profound effect on the binding of proteins which distort DNA in the DNA–protein complex. A good example of this is the nucleosome where 146 bp of DNA are wrapped in 1.8 turns around the histone protein octamer. Although the nucleosome, by its very nature, can be regarded as non-specific in terms of DNA sequence, experiments have shown that there are preferences for certain sequence motifs within the 146 bp region. These observations are likely to reflect the anisotropic flexibility of certain DNA sequences (see Chapter 3, Section 5.4).

In addition to bending flexibility, DNA also exhibits torsional flexibility, that is, flexibility resulting from a twisting motion about the DNA axis, leading to variations in the helical repeat of the DNA (Section 2.2). This type of motion is important with respect to DNA supercoiling, and is considered further in Chapter 2.

5. Conclusions

DNA is often thought of as being a rather rigid, uniform structure (possibly even as boring!). However, experimental work has revealed great structural diversity in DNA, in terms of different double-helical forms and other conformational states. As the following chapters will show, when considered in terms of its higher order structure, DNA displays a further, and even more diverse, range of conformational variations.

6. Further reading

Crothers, D.M., Haran, T.E., and Nadeau, J.G. (1990) Intrinsically bent DNA. *J. Biol. Chem.*, **265**, 7093.

Dickerson,R.E. (1983) The DNA helix and how it is read. *Sci. Am.*, **149 (Dec)**, 86.

Dickerson,R.E., Drew,H.R., Conner,B.N., Wing,R.M., Fratini,A.V., and Kopka,M.L. (1982) The anatomy of A-, B- and Z-DNA. *Science*, **216**, 475.

Drew,H.R. and McCall,M.J. (1988) Recent studies of DNA in the crystal. *Annu. Rev. Cell Biol.*, **4**, 1.

Hagerman,P.J. (1988) DNA flexibility. *Annu. Rev. Biophys. Biophys. Chem.*, **17**, 265.

Hagerman,P.J. (1990) Sequence-directed curvature of DNA. *Annu. Rev. Biochem.*, **59**, 755.

Lilley,D.M.J. (1984) DNA: sequence, structure and supercoiling. *Biochem. Soc. Trans.*, **12**, 127.

Saenger,W. (1984) *Principles of nucleic acid structure*. Springer-Verlag, New York.

Stahl,F.W. (1987) Genetic recombination. *Sci. Am.*, **56 (Feb)**, 52.

Trifonov,E.N. (1991) DNA in profile. *Trends Biochem. Sci.*, **16**, 467.

7. References

1. Watson,J.D. and Crick,F.H.C. (1953) *Nature*, **171**, 373.
2. Dickerson,R.E. (1989) *Nucleic Acids Res.*, **17**, 1797.
3. Wing,R., Drew,H., Takano,T., Broka,C., Tanaka,S., Itakura,K., and Dickerson, R.E. (1980) *Nature*, **287**, 755.
4. Saenger,W. (1984) *Principles of nucleic acid structure*. Springer-Verlag, New York.
5. Franklin,R.E. and Gosling,R. (1953) *Nature*, **171**, 740.
6. Shakked,Z., Rabinovich,D., Cruse,W.B.T., Egert,E., Kennard,O., Sala,G., Salisbury,S.A., and Viswamitra,M.A. (1981) *Proc. R. Soc. Lond.*, **B213**, 479.
7. Fairall,L., Martin,S., and Rhodes,D. (1989) *EMBO J.*, **8**, 1809.
8. Conner,B.N., Takano,T., Itakura,K., and Dickerson,R.E. (1981) *Nature*, **295**, 294.
9. Wang,A.H.-J., Quigley,G.J., Kolpak,F.J., Crawford,J.L., van Boom,J.H., van der Marel,G., and Rich,A. (1979) *Nature*, **282**, 680.
10. Drew,H.R., Takano,T., Tanaka,T., Itakura,K., and Dickerson,R.E. (1980) *Nature*, **286**, 567.
11. Pohl,F.M. and Jovin,T.M. (1972) *J. Mol. Biol.*, **67**, 375.
12. Sundquist,W.I. (1991) In *Nucleic acids in molecular biology*. Eckstein,F. and Lilley,D.M.J. (ed.), Springer-Verlag, Berlin, p. 1.
13. Wilson,J.H. (1979) *Proc. Natl. Acad. Sci. USA*, **76**, 3641.
14. Lilley,D.M.J. (1980) *Proc. Natl. Acad. Sci. USA*, **77**, 6468.
15. Lilley,D.M.J. and Hallam,L.R. (1984) *J. Mol. Biol.*, **180**, 179.
16. Gellert,M., O'Dea,M.H., and Mizuuchi,K. (1983) *Proc. Natl. Acad. Sci. USA*, **80**, 5545.
17. Courey,A.J. and Wang,J.C. (1983) *Cell*, **33**, 817.
18. West,S.C. (1992) *Annu. Rev. Biochem.*, **61**, 603.
19. Duckett,D.R., Murchie,A.I.H., Diekmann,S., von Kitzing,E., Kemper,B., and Lilley,D.M.J. (1988) *Cell*, **55**, 79.
20. Hentshel,C.C. (1982) *Nature*, **295**, 714.
21. Johnson,B.H. (1988) *Science*, **241**, 1800.
22. Htun,H. and Dahlburg,J.E. (1989) *Science*, **243**, 1571.
23. Marini,J.C., Levene,S.D., Crothers,D.M., and Englund,P.T. (1982) *Cold Spring Harbor Symp. Quant. Biol.*, **47**, 279.
24. Ulanovsky,L., Bodner,M., Trifonov,E.N., and Choder,M. (1986) *Proc. Natl. Acad. Sci. USA*, **83**, 862.
25. Wu,H.-M. and Crothers,D.M. (1984) *Nature*, **308**, 509.

26. Hagerman,P.J. (1985) *Biochemistry*, **24**, 7033.
27. Koudelka,G.B., Harbury,P., Harrison,S.C., and Ptashne,M. (1988) *Proc. Natl. Acad. Sci. USA*, **85**, 4633.
28. de Crombrugghe,B., Busby,S., and Buc,H. (1984) *Science*, **224**, 831.
29. Schultz,S.C., Shields,G.C., and Steitz,T.A. (1991) *Science*, **253**, 1001.
30. Gartenberg,M.R. and Crothers,D.M. (1988) *Nature*, **333**, 824.
31. Gartenberg,M.R. and Crothers,D.M. (1991) *J. Mol. Bol.*, **206**, 41.
32. Armstrong,F.B. (1989) *Biochemistry*. 3rd Edition. Oxford University Press, New York.

2

DNA supercoiling

1. Introduction

In the first chapter, the classical double-helical structure of DNA, and the local alterations of that structure which can occur, largely as a consequence of specific base sequences, have been described. You should now be familiar with these concepts. The main subject of this book, namely the supercoiling of DNA, a global alteration of structure which arises directly from the double-helical nature of DNA, will now be addressed.

2. Historical perspective

2.1. Closed-circular DNA and supercoiling

In the early 1960s, a number of groups were interested in the DNA molecules of viruses, in particular the DNA tumour virus, polyoma. During sedimentation analysis, which separates molecules according to size and compactness, DNA from polyoma virus consistently fractionated into two components, I and II. Both components were shown to comprise double-stranded DNA of the same molecular weight. However, the major component (I) had a higher sedimentation co-efficient (i.e. was more compact), and was unusually resistant to denaturation (the melting of base-pairing between the strands; see Chapter 1, Section 2.1) on heating or exposure to high pH. Furthermore, following denaturation, the single strands did not separate from each other. This led to the suggestion (1,2) that component I comprised 'circular base-paired duplex molecules without chain ends', and component II was the linear form of the same molecule. In other words, the DNA molecules in component I consist of two anti-parallel circular DNA single-strands winding helically around one another, that is, the ends of each strand of a linear double-stranded molecule are joined in the conventional 5′ to 3′ manner to form a double-stranded circle (*Figure 2.1*).

This interpretation was supported when electron micrographs of polyoma DNA showed a preponderance of circular molecules. However, a problem arose

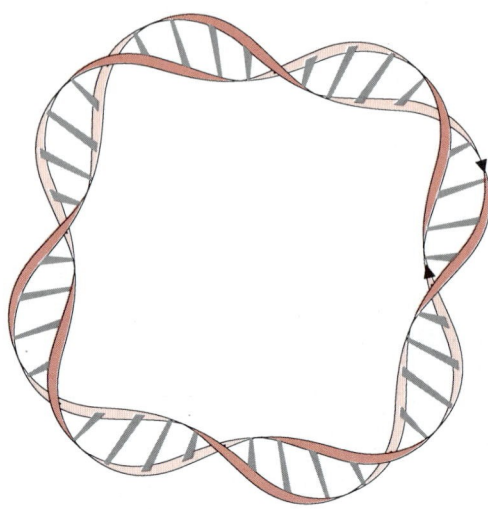

Figure 2.1. A schematic representation of a closed-circular DNA molecule. Each of the single strands of a double-stranded DNA double helix is covalently closed via a phospho-diester bond to form a circular double-stranded molecule.

with the discovery that a single break in one of the two strands of component I, caused by the endonuclease DNase I, converted it directly to component II. How could the cleavage of one strand of a closed-circular duplex convert it to a linear molecule? Subsequently, electron micrographs of the component II molecules formed in this way showed them also to be circular. So, how could two circular DNA molecules, I and II, distinguished only by the breakage of one backbone phosphodiester bond, have such different properties?

The clue to the resolution of this conundrum came from the electron micrographs. Electron micrographs of the component I molecules showed many crossings of the DNA double strands (*Figure 2.2a*), whereas the component II molecules were mainly open rings (*Figure 2.2b*). Such crossings had previously been tentatively ascribed to protein cross-links, but Jerome Vinograd and his colleagues (3) suggested that such crossings could represent a 'twisted circular form' of the DNA, which explained the more compact nature of component I. Such a conformation would result if, conceptually, before the joining of the ends of a linear duplex into a closed-circular molecule, one end was twisted relative to the other, thus introducing some strain into the molecule. This is analogous to the behaviour of any length of elastic material; such a relative twisting of the ends tends to be manifested as a coiling of the material upon itself. One of the best models of this behaviour of DNA is a length of rubber tubing. If the unconstrained tubing represents the double helix of a linear DNA, then a relative twisting of the ends, followed by the closure of the tubing into a circle with a connector, will result, when external constraints are released, in something like *Figure 2.3*. Such a coiling of the DNA helix upon itself is the literal meaning of supercoiling;

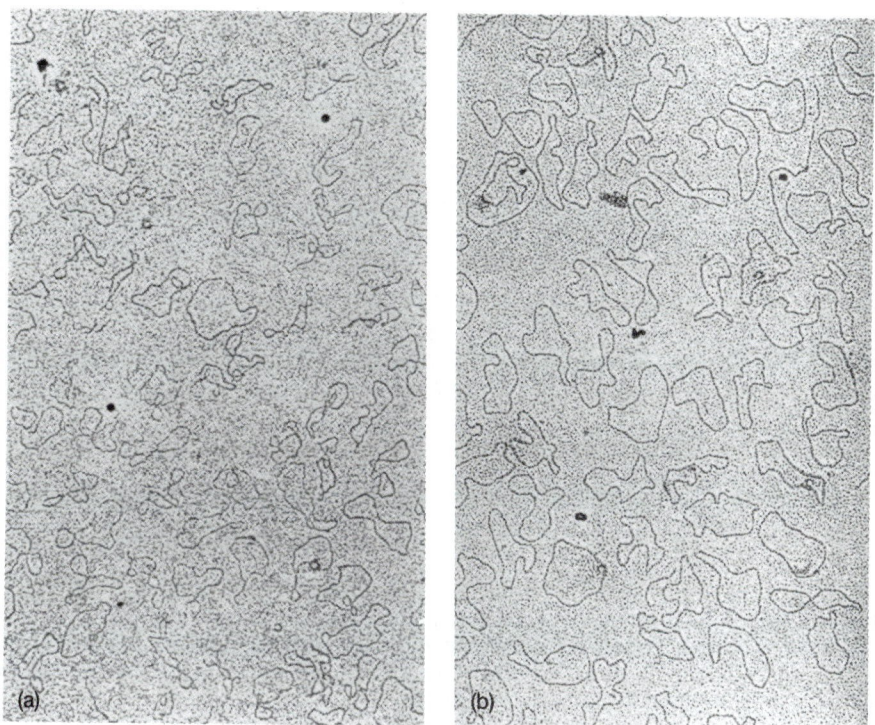

Figure 2.2. Electron micrographs of DNA isolated from the polyoma virus. **(a)** Component I, **(b)** component II (reproduced from ref. 3).

that is a higher order coiling of the DNA helix. Furthermore, such supercoiling is locked into the system. If the two DNA strands are joined covalently across the original break 5'–3', the elastic strain which results in superhelicity cannot be released without the breakage of one or both strands. This 'strand breakage' corresponds in the tubing model to a reversal of the twisting at the break point. The DNA molecule may be geometrically contorted in a variety of ways, as is described later, but the basic strained state of the molecule cannot be changed without strand breakage.

It can now be seen how the cleavage of one phosphodiester bond of the polyoma supercoiled DNA (I) leads directly to the open-circular, unconstrained component II, since the broken strand can rotate about the intact strand to dissipate the torsional stress. Component II is often known as 'open-circular' or 'nicked-circular' DNA, referring to the single break in one strand. The closure of a linear DNA into a planar circle without further constraint leads to 'relaxed' closed-circular DNA. Although nicked-circular and relaxed DNA have a similar conformation, they differ chemically by the presence or absence of one single-strand break.

If the reader has an intuitive feel for these properties of rubber tubing, and by analogy, of DNA, then read on. If not, make a rubber tubing model and try the

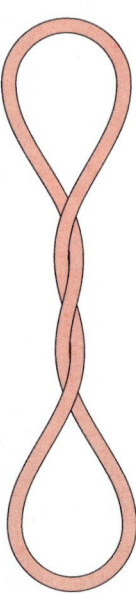

Figure 2.3. A tubing model of supercoiled DNA. A length of rubber tubing closed by a connector will adopt a conformation similar to that shown if the ends are twisted relative to each other before closure. If a double helix is imagined drawn on the tubing, this is quite a good model for the behaviour of DNA.

exercise above. This may seem to be labouring a small point, but without a firm understanding of these principles, your ability to follow the remainder of this discussion will rapidly deteriorate!

It may not have escaped your attention that the superhelical turns described may be of either handedness, corresponding to the introduction of right- or left-handed twisting into the linear molecule before circle formation. Since double-stranded DNA is a right-handed helix, right-handed twisting tends to twist up the helix further, and left-handed twisting leads to an untwisting of the helix. The two opposite forms of supercoiling introduced in this way are designated positive [for twisting in the same direction as the DNA helix (right-handed)] and negative (in the opposite, left-handed direction). Vinograd correctly deduced the handedness of the supercoiling of polyoma component I DNA to be negative.

Thus, with the powerful, but (in retrospect) simple insight that a closed-circular DNA molecule could exist in a 'twisted circular' or supercoiled form, Vinograd and his colleagues neatly explained the properties of polyoma virus DNA, and analogous results pertaining to other circular DNA molecules. However, they opened up a whole new field of study, since behind this straightforward idea is concealed a large number of theoretical and practical ramifications, ranging from the mysteries of linking number, twist and writhe (see below) to the processes required to untangle (often literally!) the biological consequences of these unexpected topological properties of the DNA molecule.

3. A quantitative measure of DNA supercoiling

As previously stated, the elastic stress associated with a supercoiled closed-circular DNA molecule cannot be relieved without the breakage of one or both backbone strands, although the molecule can be geometrically deformed by external forces. Is it possible to devise a quantitative measure of the supercoiling of a DNA molecule?

3.1. Linking number

If a linear double-stranded DNA molecule is closed into a circle (*Figure 2.1*), by the formation of 5′–3′ phosphodiester bonds to close each strand, then the strands will be linked together a number of times corresponding to the number of double-helical turns in the original linear molecule. This number, which must be an integer, is known as the linking number of the two strands, or the linking number of the closed-circular DNA molecule, abbreviated as Lk. Earlier literature uses the designation α, referred to as the topological winding number; this term is identical to Lk.

This subtle concept may be neatly demonstrated by modelling DNA as a length of ribbon, as illustrated in *Figure 2.4*, where the two edges of the ribbon represent the single strands of DNA. Initially there are no double-helical turns, and the ribbon is closed into a simple circle (*Figure 2.4a*). Hence this models a hypothetical situation where the two strands are totally unwound; the two ribbon edges are not linked. If the ribbon is now cut (*Figure 2.4b*), and rejoined after a full 360° twist has been made in one end, a model of a closed-circular DNA with one double-helical turn is obtained (*Figure 2.4c*). Remember that although a 180° twist will allow rejoining of the ribbon, in the case of DNA, this will place two 3′ and two 5′ ends together, and hence is not allowed. The linking of the two strands (edges) can now be shown by cutting longitudinally around the ribbon (*Figure 2.4c*), which models the separation of the two strands, to yield two circles linked together once (*Figure 2.4d*); that is with a linking number of 1 ($Lk = 1$). It is well worth making such a model to convince yourself of this phenomenon; in the absence of ribbon, use a long strip of paper and tape. In the same way, it can be demonstrated that the introduction of n turns into the ribbon, corresponding to n double-helical DNA turns, leads to two circles linked n times. If n gets too large, the model becomes confusing, but colouring one edge initially will help you keep track.

Of course, a given length of real DNA has an inherent number of double-helical turns by virtue of its structure. This number is the length of the DNA, N, in base pairs (usually thousands for natural DNAs) divided by the number of base pairs per turn of helix, h, which is dependent on the conditions, although a standard value, $h°$, is defined under standard conditions [0.2 M NaCl, pH 7, 37°C; (4)] and is often taken to be 10.5 bp/turn (see Chapter 1, Section 2.2). The value of N/h will not in general be an integer, so when the DNA is bent into a simple, planar circle, the strand ends will not line up precisely, although the slight twisting required to join the ends is insignificant over thousands of base pairs. Hence a DNA molecule joined into a circle with the minimum of torsional stress will have a

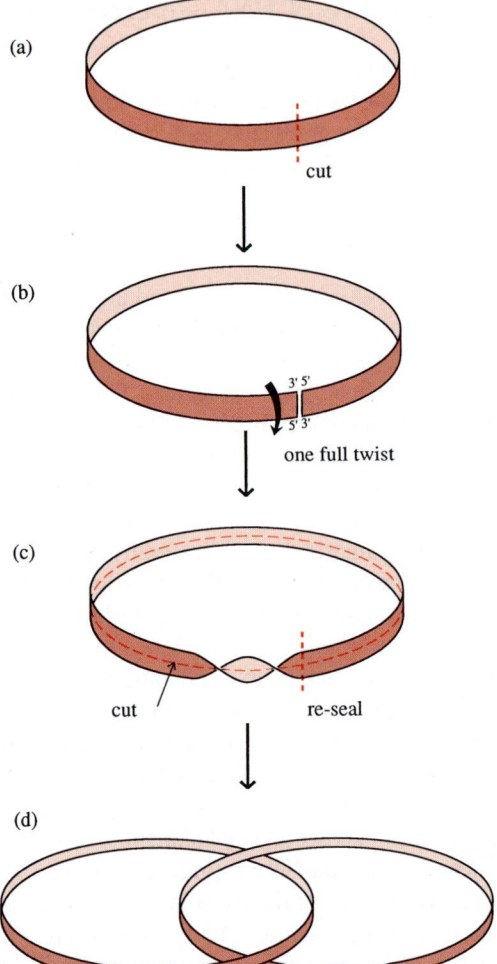

Figure 2.4. Demonstration of linking number (Lk), using a length of ribbon or paper. **(a)** A length of ribbon is taped into a circle. Each edge of the ribbon represents one strand of DNA. The edges (strands) are initially unlinked. **(b)** The ribbon is cut and a 360° turn is introduced into the ribbon, mimicking one turn of a DNA double helix. **(c)** The ribbon is rejoined, then cut longitudinally, to model the separation of the strands. **(d)** This produces two ribbons (single strands), linked together once.

linking number which is the closest integer to N/h. This 'standard' linking number is referred to as Lk_m (5); hence:

$$Lk_m \approx \frac{N}{h} \approx \frac{N}{10.5} \qquad (2.1)$$

The linking number of the strands of right-handed DNA is conventionally taken to be positive. Thus, for example, for plasmid pBR322, with 4361 bp (N), Lk_m is +415, under conditions where the helical repeat (h) is 10.5 bp/turn.

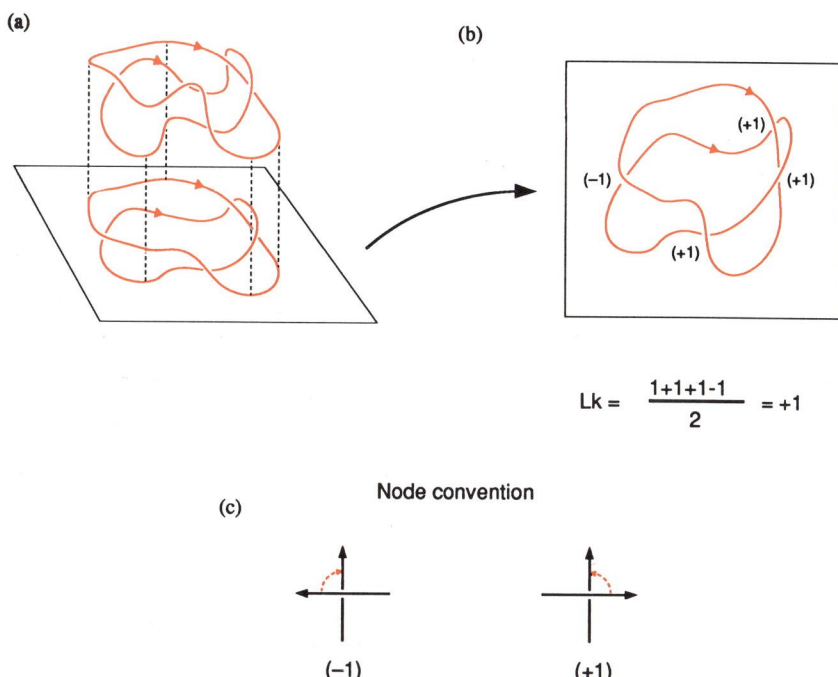

$$Lk = \frac{1+1+1-1}{2} = +1$$

Node convention

(c)

(−1)   (+1)

Figure 2.5. The definition of linking number (*Lk*). **(a)** Two closed curves are projected onto a plane (this can be in any direction). **(b)** The curves are each assigned a polarity, and the crossings of one curve over the other (nodes) are given a number (±1) according to the convention in **(c)**. Self-crossings of a curve do not count. The linking number is equal to half the sum of the node numbers.

3.1.1. A strict definition of linking number

Linking number turns out to be a general property of any two closed curves in three-dimensional space. Rigorously, the linking number may be determined using the number and handedness of the cross-overs of the two curves when projected onto a plane. Projection onto a plane is illustrated in *Figure 2.5a*, and may be thought of as a two-dimensional representation of the three-dimensional curves viewed from any direction, except that the upper and lower strands must be distinguishable at the cross-overs, also called 'nodes'. The curves are each given a direction, and a value of +1 or −1 assigned to each crossing of one curve over the other, according to the direction of movement required to align the upper with the lower strand, with a rotation of less than 180°; clockwise being negative and counter-clockwise positive (confusing, but illustrated in *Figure 2.5c*). The sum of the values assigned is halved to give the linking number of the two curves (*Figure 2.5b*). This division by two makes sense, since a single link (e.g. *Figure 2.6a*) requires two crossings of the same handedness. This method works for any projection of two curves, and takes account of any fortuitous overlapping of the curves which does not constitute a link, for example in the left-hand half of *Figure 2.5b* the signs of the two nodes produced cancel out. Likewise, this definition helps to show that linking number is unchanged without

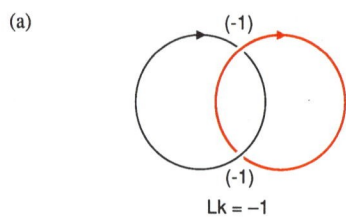

(a)

(-1)

(-1)

Lk = −1

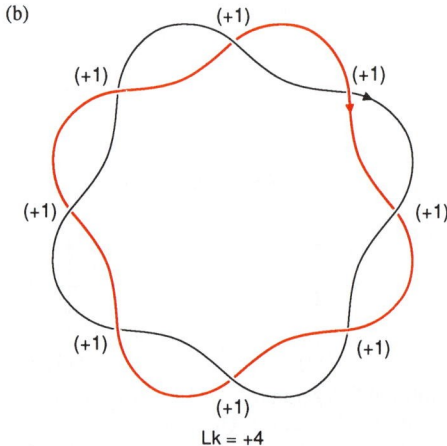

(b)

(+1)

(+1)

(+1)

(+1)

(+1)

(+1)

(+1)

(+1)

Lk = +4

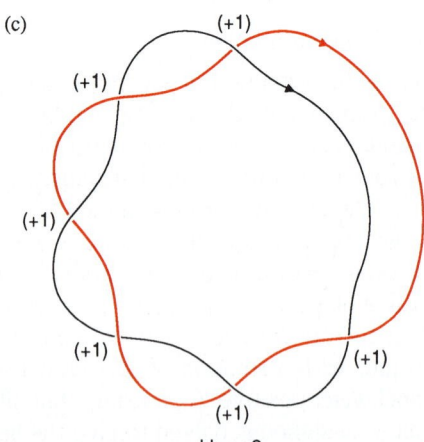

(c)

(+1)

(+1)

(+1)

(+1)

(+1)

(+1)

Lk = +3

Figure 2.6. Linking number examples. **(a)** Two singly-linked curves. In this case $Lk = -1$; the sign is determined by the directions assigned to the curves. **(b)** A hypothetical DNA molecule; the strands must be assumed to run in the same direction for the linking number to be positive. **(c)** If a DNA molecule is untwisted by one turn before closure into a circle, the linking number decreases by one.

breakage of one of the curves. Any stretching or bending (technically, smooth deformation) of the curves may add or subtract nodes in a given two-dimensional projection, but only in self-cancelling positive and negative pairs, leaving Lk unchanged. The application of this method to a model DNA circle is shown in *Figure 2.6b*; by convention, the strands are shown running in the same direction, to ensure that the resulting Lk is positive. Of course, in a structural sense, the strands of DNA run anti-parallel.

So, the linking number (Lk) is a measure of a fundamental property of a closed-circular DNA molecule, which has no meaning until both strands are sealed, and is unaffected by changes in the conformation of the DNA, as long as these do not involve the breakage of one or both strands. Linking number is thus a *topological* property of closed-circular DNA, and does not depend on the particular *geometry* of the DNA. DNA molecules differing only in linking number are known as topological isomers, or topoisomers.

3.2. Supercoiling, linking difference, and specific linking difference

It can now be seen that the process of introducing supercoiling into a DNA molecule (relative twisting or untwisting of the right-handed helix before closure of a linear molecule into a circle) reduces or increases the number of helical turns trapped by that closure and results in (necessarily integral) changes in the linking number of the final closed-circular species (*Figure 2.6c*). A change in supercoiling in a given molecule is thus accompanied by a change in the linking number. Specifically, the change in linking number from Lk_m, which corresponds to the most unconstrained closed-circular (relaxed) topoisomer as discussed above, is a measure of the extent of supercoiling. This value, $Lk - Lk_m$, is known as the linking difference, or ΔLk, and may be positive or negative. Because the linking number of closed-circular B-DNA is defined as positive, twisting up of the helix before closure leads to an increase in linking number above Lk_m, that is a positive linking difference, and corresponds to positive supercoiling. Analogously, unwinding of the helix before closure yields a negative ΔLk and negative supercoiling.

Strictly speaking, the linking difference (ΔLk) of a supercoiled DNA molecule should be measured from the hypothetically unconstrained state corresponding to the closure of a linear molecule into a planar circle without *any* torsional strain. This is because the small twisting which may be required to form the Lk_m isomer (Section 3.1) should really be counted towards the total supercoiling of the molecule. This 'hypothetical' linking number, the number of double-helical turns in the original linear molecule is exactly equal to N/h (cf. equation 2.1), and need not be an integer. This quantity is known as $Lk°$. $Lk°$ is not a true linking number, since a linking number must be integral, but serves as a reference point for the measurement of the level of supercoiling. Thus a better general definition of linking difference is:

$$\Delta Lk = Lk - Lk° \qquad (2.2)$$

For moderate supercoiling in plasmid-sized DNA molecules ($N >$ a few thousand base pairs), the distinction between Lk_m and $Lk°$ is not too significant; for example, $Lk°$ for pBR322 is 415.3, if $h = 10.5$ bp/turn (cf. Section 3.1). The physical relevance of $Lk°$ will be discussed in Section 5.2.

As has been described, the introduction of supercoils into a DNA molecule corresponds to the introduction of torsional stress. A given ΔLk will produce more torsional stress in a small DNA molecule than in a large one; hence linking difference is commonly normalized to the size of the DNA circle, by dividing by $Lk°$, to give the specific linking difference (σ):

$$\sigma = \frac{Lk - Lk°}{Lk°} = \frac{\Delta Lk}{Lk°} \tag{2.3}$$

This gives a measure of the extent of supercoiling which can be used for comparisons between DNA molecules. For example, natural closed-circular DNA molecules, such as bacterial plasmids, or the *Escherichia coli* chromosome, while varying widely in size, have, when isolated, a specific linking difference of around -0.06. Specific linking difference is often referred to as superhelix or superhelical density; however, these terms potentially confuse topological and geometric properties, and should probably be avoided (see Section 4.3).

4. Geometric properties of closed-circular DNA

4.1. Twist and writhe

A means of expressing the geometric conformations which may be adopted by closed-circular DNA in response to changes of supercoiling, or ΔLk, is now required. As previously described, one way of imagining the generation of DNA molecules of varying Lk is to envisage a twisting up or untwisting of a linear DNA helix, before its closure into a double-stranded circle. This implies that one possible geometric consequence of a change in linking number is a corresponding change in the winding of the DNA double helix itself. In the case of negatively supercoiled DNA, this corresponds to an untwisting of the DNA helix, and an increase in the helical repeat, h (number of bp/turn).

On the other hand, using the rubber tubing model, it can be seen that a change in linking number from the most relaxed (Lk_m) topoisomer leads to a coiling of the tubing upon itself (*Figure 2.3*). This is the behaviour of DNA which first alerted Vinograd to the supercoiling phenomenon, as recounted above. In fact, these two behaviours are complementary to each other, and each may be defined and described. The twist (Tw) describes how the individual strands of DNA coil around one another (or, more rigorously, around the axis of the DNA helix). The writhe (Wr) describes how the helix axis coils in space. Both are complex geometric functions, whose values need not be integral and may not in general be easily calculated, although, as described later in Section 4.2 and in Chapter 3, several special cases may be identified. However, the central important result for the study of DNA supercoiling is that the twist and writhe of a given DNA molecule must sum to the linking number. That is:

$$Lk = Tw + Wr \tag{2.4}$$

This implies that for a given closed-circular molecule, Lk being invariant, any change in the twist of the DNA must be accompanied by an equal and opposite change in the writhe, and vice-versa. Furthermore, any change in linking number manifests itself geometrically as a change in the twist and/or the writhe:

$$\Delta Lk = \Delta Tw + \Delta Wr \qquad (2.5)$$

These results were originally proved as a piece of pure mathematics, without reference to DNA, by James White in 1969 (6), and were later independently developed by F. Brock Fuller (7,8) as a rigorous examination of the results of Vinograd and colleagues. More accessible discussions (for the biologist) of the technical aspects of this area are given in the references (9–11).

4.2. The interconversion of twist and writhe

The best way to get a feel for the properties of twist and writhe is to consider the behaviour of DNA as modelled by rubber tubing. The first case to consider is when the tubing (or DNA) is closed into a planar circle (*Figure 2.7a*). It can be shown, and indeed it seems intuitive, that the writhe of a plane circle is zero. Hence, in this case, from Equation 2.4, the twist of a DNA is equal to its linking number (i.e. when $Wr = 0$, $Lk = Tw$), that is, the twist is equal to the number of double-helical turns around the circle; this makes good sense.

The identity $Lk = Tw$ is only normally true if the helix axis lies in the plane. This will not generally be the case for the most relaxed topoisomer with $Lk = Lk_m$. As seen in Section 3.1, the strand-ends of a linear DNA will not normally line up precisely, since N/h will not usually be an integer, and the slight distortion required to close the circle may be manifested as a small writhe. It is therefore useful to define the term $Tw°$, which is the twist of the linear DNA before ring closure, or the twist of the nicked-circular (open-circular) form of the DNA. That is:

$$Tw° = \frac{N}{h} \qquad (2.6)$$

This is identical to the previous definition of $Lk°$ (Section 3.2). It is important to note, however, that $Lk°$ was defined for a planar circle. If the unconstrained DNA does not lie in a plane, for example because it contains regions of intrinsic curvature (Chapter 1, Section 4.1), then $Lk°$ may contain a contribution from the writhe corresponding to this lack of planarity. In this special case, $Lk° \neq Tw°$; $Lk° = Tw° + Wr°$ (cf. Equation 2.4), where $Wr°$ is the writhing contributed by the intrinsic structure of the DNA, rather than by any deformation due to supercoiling. It should be pointed out in this context that there is at least one special case where the writhe of a non-planar DNA is equal to zero. If the DNA axis can be made to lie on the surface of a sphere, without intersection, then $Wr = 0$ and $Lk = Tw$.

It is helpful to draw a line on the relaxed rubber tubing, as in *Figure 2.7a* as a marker to follow changes in the twist of the tubing. If the tubing is broken, and one end rotated four times in a left-handed sense before reclosure (the line must be continuous across the break to ensure integral turns), the tubing now models a DNA with $\Delta Lk = -4$. The model will now adopt the interwound, or 'plectonemic'

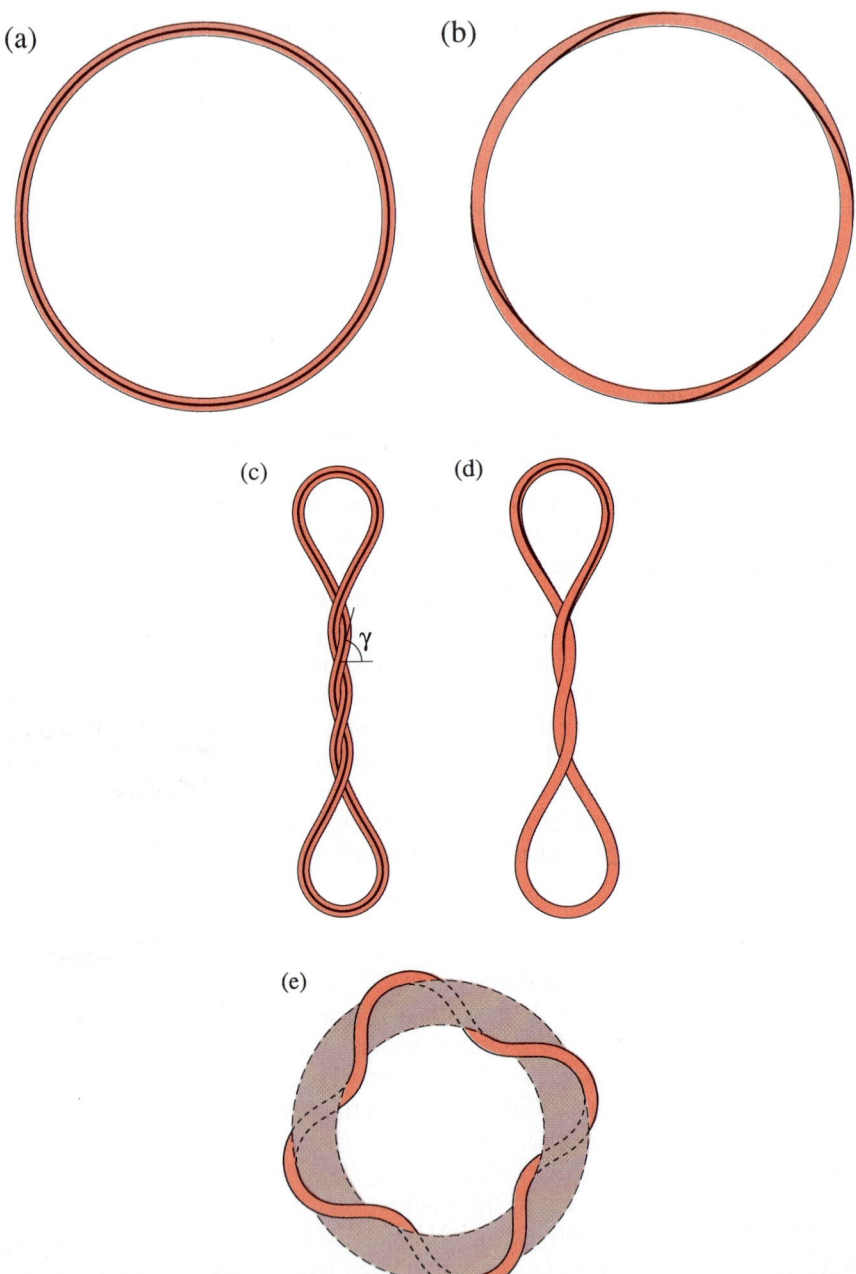

Figure 2.7. Twisting and writhing in the tubing model of DNA. **(a)** 'Relaxed' DNA; Lk = Lk_m. **(b)** Untwisted DNA; $\Delta Lk = -4$, $\Delta Tw = -4$, $Wr = 0$. **(c)** Plectonemic, or interwound conformation of supercoiled DNA; $\Delta Lk = -4$, $\Delta Tw \approx 0$, $Wr \approx -4$. γ is the superhelix winding angle (see text). **(d)** The plectonemic conformation with lowest energy; $\Delta Lk = -4$, $\Delta Tw \approx -1$, $Wr \approx -3$. **(e)** The 'toroidal' winding of DNA, in which the axis of the DNA helix lies on a real or imaginary torus (shaded region).

conformation previously described (*Figure 2.3*), in which the helix axis is considerably contorted. By deformation, two extreme cases may now be demonstrated. If the tubing is constrained to lie in a plane, it can be seen that a twisting of the helix is required to accomplish this movement. $Wr = 0$ again, and hence $\Delta Tw = -4$ (Equation 2.5). These twists may be counted as the four left-handed rotations of the marker line around the tubing (*Figure 2.7b*). Thus the linking difference is manifested as an untwisting of the DNA double helix.

Alternatively, the molecule may be deformed to a conformation shown in *Figure 2.7c*. For a helix whose axis does not lie in a plane, the contributions of twist and writhe are not always easy to determine; however for a regular right-handed plectonemic superhelix, the writhe is equal to $-n \sin \gamma$, where n is the number of nodes (4 in this case) and γ is the superhelix winding angle (the pitch angle of the plectonemic superhelix, illustrated in *Figure 2.7c* (11)). For an extended superhelix of this sort, where γ approaches $90°$, $Wr \approx -n \approx -4$. From the continuous presence of the marker line on the upper surface of the tubing, it can also be seen that the ΔTw of the tubing is close to zero; hence $\Delta Lk = -4$; $\Delta Tw \approx 0$; $\Delta Wr = Wr \approx -4$. The most stable conformation (*Figure 2.7d*) will probably involve both changes in twist and writhe, and lies between the two extremes shown in *Figure 2.7b* and *c*; the contribution of twist and writhe to this conformation reflects the relative unfavourability of untwisting the DNA helix (Tw) and bending the helix axis (Wr).

4.2.1. Plectonemic and toroidal conformations

The conformation considered up until now has been the plectonemic, or interwound conformation, in which the DNA helix is wound around another part of the same molecule in a higher order helix (*Figure 2.7c* and *d*). The sense of this interwinding is right-handed in the case of negatively supercoiled DNA. However, it is also necessary to consider an alternative conformation, toroidal winding, so called because the axis of the DNA helix lies on the surface of an imaginary torus (*Figure 2.7e*). This conformation most closely corresponds to the literal meaning of the term 'superhelix'; that is, a left-handed untwisting of the DNA helix is manifested as a left-handed helix of a higher order wound around the torus (cf. the right-handed superhelix in the interwound form). Unfortunately, an attempt to produce toroidal winding with the rubber tubing model does not work very well in the absence of a real torus to constrain the tubing; writhing of the tubing axis is achieved with less severe bending in an interwound superhelix. The available evidence suggests that the same is true for free supercoiled DNA both *in vitro* and *in vivo*. Recent measurements of supercoiled DNA molecules in electron micrographs (12,13) have revealed only interwound molecules and have suggested that moderate levels of ΔLk are partitioned into approximately 75% ΔWr and 25% ΔTw. Studies of the Int recombination reaction (see Chapter 6, Section 6.1) have concluded that plectonemic supercoiling is most likely to be the conformation present *in vivo*. Despite these results, the toroidal conformation still merits consideration, since it is the model for the binding of DNA on many protein surfaces (see Section 5.4 and Chapter 3, Section 5).

Toroidal winding of DNA may be conveniently illustrated by considering a coiled telephone wire (*Figure 2.8a*). This has substantial writhe, but very little twist. Stretching out of the telephone wire causes an interconversion of the writhe to twist (*Figure 2.8b*); the ultimate endpoint is a highly twisted linear wire (*Wr* = 0; *Figure 2.8c*). Although a telephone wire is not a model of a closed-circular DNA, it is still meaningful to consider its twist and writhe. Twist and writhe are geometric, rather than topological, properties, and may be defined for sections of circular DNA or linear DNA.

4.3. Nomenclature

Just a few words about the nomenclature used to describe the phenomenon of DNA supercoiling. The term supercoiling itself was coined to describe the coiling of the DNA helix axis upon itself; likewise superhelix (helix of helices). As already described, however, the topological extent of supercoiling (ΔLk) is in general not manifested only as a 'supercoil', but also as an alteration in the DNA helix parameters (unwinding or untwisting). Expressions such as extent of supercoiling, number of supercoils, and supercoil or superhelix density could ambiguously refer to either changes in linking number or changes in the writhing component of ΔLk. Such terms are therefore probably best avoided in a technical discussion, although supercoiling is acceptable as a vernacular description of the overall process.

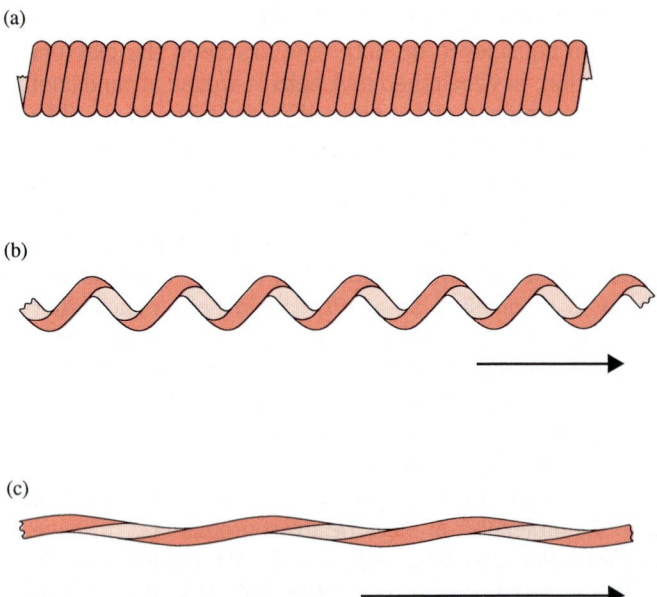

(a)

(b)

(c)

Figure 2.8. A telephone wire as a model of toroidal writhing of DNA. **(a)** The un-stretched wire has a lot of writhe, but little twist. **(b)** As the wire is stretched, writhe is converted to twisting of the wire. **(c)** Eventually, the wire is almost straight (unwrithed), but highly twisted.

Before the introduction of the rigorously defined geometric parameters, twist and writhe, another nomenclature was devised for the description of supercoiled DNA by Vinograd and colleagues. The linking number of a closed-circular DNA, then known as α, was made up of the sum of β, the 'duplex winding number' and τ, the number of 'superhelical turns'; that is:

$$\alpha = \beta + \tau \qquad (2.7)$$

These terms suffer from a similar imprecision to that described above. Some usages imply that β is equivalent to twist, and τ to writhe (14), whereas in others β seems most similar to Tw° (or Lk°) and τ to ΔLk (15). Therefore, although these terms appear in much of the early literature, it is important to try to recast the arguments in terms of twist and writhe. Although $\alpha = Lk$, it is *not* safe to assume that $\beta = Tw$ and $\tau = Wr$. For a more detailed discussion of the relationship between these quantities, see Cozzarelli *et al.* (11).

5. Topology and geometry of real DNA

The previous discussion of topological and geometric properties of closed-circular rubber tubing has perhaps been rather abstract, so the discussion will now centre on real DNA and consider some of the methods which can be used to analyse supercoiling.

Almost all DNAs in nature, from small bacterial plasmids to huge eukaryotic chromosomes exhibit the properties described in the previous sections. Furthermore, most DNAs *in vivo* are negatively supercoiled, that is they have a negative linking difference ($Lk < Lk^\circ$). The ubiquity of supercoiling of DNA *in vivo* will be discussed in more detail in Chapter 6.

5.1. Agarose gel electrophoresis of plasmids

The molecules most commonly used in the investigation of the topological properties of DNA are bacterial plasmids of a few thousand base pairs in length. The most common method of analysis is agarose gel electrophoresis (16,17). Electrophoresis separates DNA molecules on the basis of size and compactness; smaller and/or more compact molecules will migrate more rapidly through the matrix of the gel under the influence of the electric field.

A typical negatively supercoiled plasmid, for example the *E. coli* plasmid pBR322 ($N = 4361$ bp), separates into two bands on an agarose gel (*Figure 2.9a*). The band of higher mobility corresponds to negatively supercoiled molecules, having the compact, writhed structure modelled in *Figure 2.3*. The lower mobility band comprises nicked-circular (open-circular) molecules, having the less compact planar circle conformation. Such nicked-circular molecules are not necessarily present *in vivo*, but may be formed by breakage of one strand of the closed-circular molecules during purification. These supercoiled and nicked bands are analogous respectively to components I and II identified by sedimentation analysis of polyoma virus DNA (see Section 2.1). The mobility of linear DNA of the same

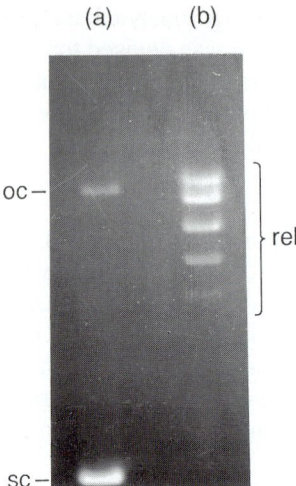

Figure 2.9. Agarose gel electrophoresis of closed-circular pBR322 DNA. **(a)** Super-coiled DNA isolated from bacteria commonly migrates as two bands on an agarose gel. sc; negatively supercoiled (closed-circular) pBR322 DNA, oc; open circular (nicked) pBR322 DNA. **(b)** The same material as in **(a)**, after treatment with a topoisomerase, consists of a distribution of topoisomers. rel; relaxed pBR322 DNA.

molecular weight varies depending on the conditions of electrophoresis; how-ever, it is most often found to migrate between the nicked and supercoiled species.

5.2. Relaxation with topoisomerases

Negatively supercoiled DNA has a higher free energy than its relaxed equivalent, due to its torsional stress. A class of enzymes, known as topoisomerases due to their ability to alter the topological state of DNA, are able to relieve the torsional stress of supercoiled DNA, by a mechanism necessarily requiring the transient breakage of one or both DNA strands (see Chapter 5). The product of the action of such an enzyme is relaxed DNA, that is, the linking number of the DNA approaches $Lk°$. Treatment of closed-circular DNA with a topoisomerase in this way is equivalent to the closure of a linear or a nicked-circular molecule into a closed-circle by ligation under the same conditions (14). The result of electro-phoresis of such a product on an agarose gel is shown in *Figure 2.9b*. An interesting property of relaxed DNA is that it migrates as a series of bands rather than a single band (the nicked-circular band is still there, unchanged by the enzyme). This is because a topoisomerase catalyses equilibration between topoisomers (DNAs of different linking number). The energy difference between molecules of similar linking number around $Lk°$ (relaxed DNA) is less than the thermal energy available at normal temperatures; hence the equilibrated sample consists of a mixture of topoisomers, with a distribution of values of Lk, centred at $Lk°$. Adjacent bands on the gel correspond to isomers differing by one in their linking number and the most abundant consists of molecules with $Lk = Lk_m$. The range of Lk values in relaxed DNA represents a Boltzmann distribution determined by thermal energy; the form of the distribution is Gaussian (see Section 6.2).

For circular molecules of the same molecular weight, migration in an electric field through an agarose gel is essentially a measure of writhe. Although adjacent bands differ in linking number, it is any associated differences in writhe which cause their different mobilities in the gel. The so called 'native' negatively supercoiled plasmid DNA isolated from bacteria (*Figure 2.9a*) also consists of a distribution of topoisomers, but above a certain (negative or positive) value of writhe, the topoisomers run as an unresolved single band of high mobility.

The average value of linking number obtained on relaxation depends on the solution conditions under which the relaxation took place. Changes in the conditions (e.g. temperature, ionic strength) alter the helical repeat (h) of the DNA, and hence change the value of $Tw°$, and hence $Lk°$ (Equation 2.6). Lk_m is the linking number of the most abundant topoisomer in the relaxed distribution, and under the conditions of relaxation will have the most energetically favourable conformation, that is, on average, close to a plane circle. The Lk_m topoisomer should thus have virtually the same mobility on a gel as the nicked-circular species, but this will only happen in the unlikely situation that electrophoresis is carried out under the same conditions as the relaxation reaction (or under conditions where, for example, ionic strength and temperature have equal and opposite effects on h). In such circumstances, the Lk_m topoisomer will have the lowest mobility and the isomers with $\Delta Lk = \pm 1$, ± 2 etc. will have steadily increasing mobilities (topoisomers of positive and negative ΔLk have positive and negative writhe respectively, but each leads to a similar increase in the compactness of the DNA). In general, however, the nominally relaxed topoisomers formed in the topoisomerase reaction are induced to adopt new conformations by the changed conditions of gel electrophoresis (*Figure 2.9b*). Specifically, the change in twist caused by the altered conditions forces the writhe to change from its initial average value of zero to a new value, and hence the DNA will run with a higher mobility in the gel.

Hence, in colloquial language, a DNA molecule that is relaxed in a given topoisomerase reaction may not be relaxed when run on an agarose gel, and in general, whether a topoisomer of a given linking number is positively supercoiled, negatively supercoiled or relaxed depends on the conditions; in other words on the value of h and hence $Tw°$ (or $Lk°$).

5.3. The effect of intercalators

An important example of a factor affecting the twist and helical repeat of a DNA is the presence of intercalating molecules. An intercalator contains a planar, usually polycyclic, aromatic structure which can insert itself between two base pairs of double-stranded DNA. This causes a local unwinding of the DNA helix, resulting in an overall increase in the helical repeat, that is, a decrease in the twist of the DNA. The classic example of an intercalating molecule is the dye, ethidium bromide (EtBr, *Figure 2.10a*), which binds tightly to double-stranded DNA.

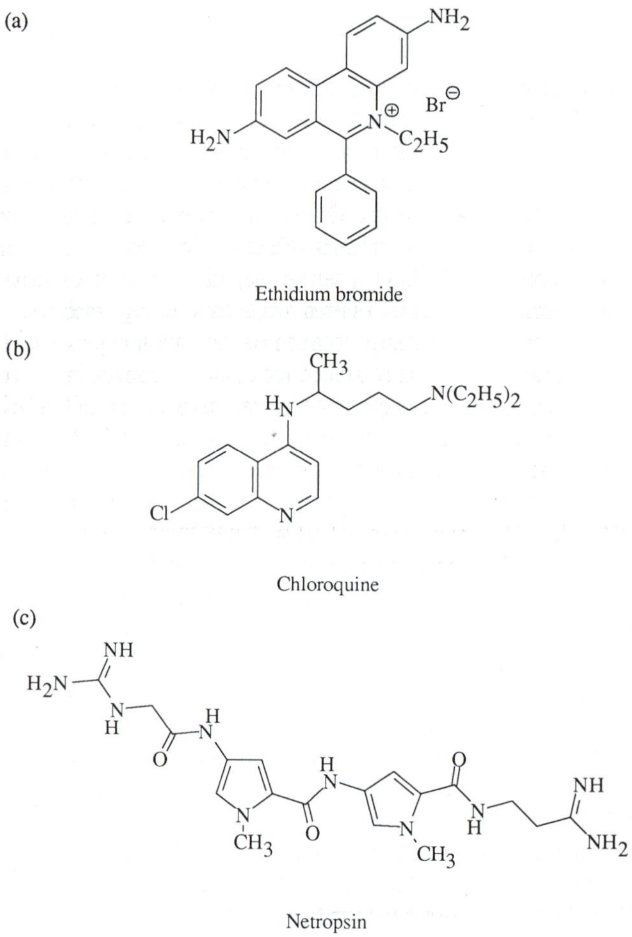

(a)

Ethidium bromide

(b)

Chloroquine

(c)

Netropsin

Figure 2.10. Structures of molecules whose binding alters the geometry of double-stranded DNA. **(a)** Ethidium bromide; **(b)** chloroquine; **(c)** netropsin.

Ethidium bromide is well known as a stain for DNA, since it exhibits a large enhancement of fluorescence on binding to DNA, but its additional interest from the point of view of DNA topology is that the binding of each molecule between adjacent base pairs causes a local unwinding of the helix of 26° (18,19).

From Equation 2.4, a decrease in the twist of a closed-circular DNA molecule results in an increase in the writhe. Hence a DNA sample run on a series of gels containing increasing concentrations of EtBr will exhibit a gradual shift towards more positive values of writhe, and the average mobility of the topoisomer distribution will change accordingly (*Figure 2.11a*). For example, if the DNA sample is relaxed on the gel under certain conditions, increasing EtBr will cause an increasing positive writhe, and higher and higher mobility, until the DNA

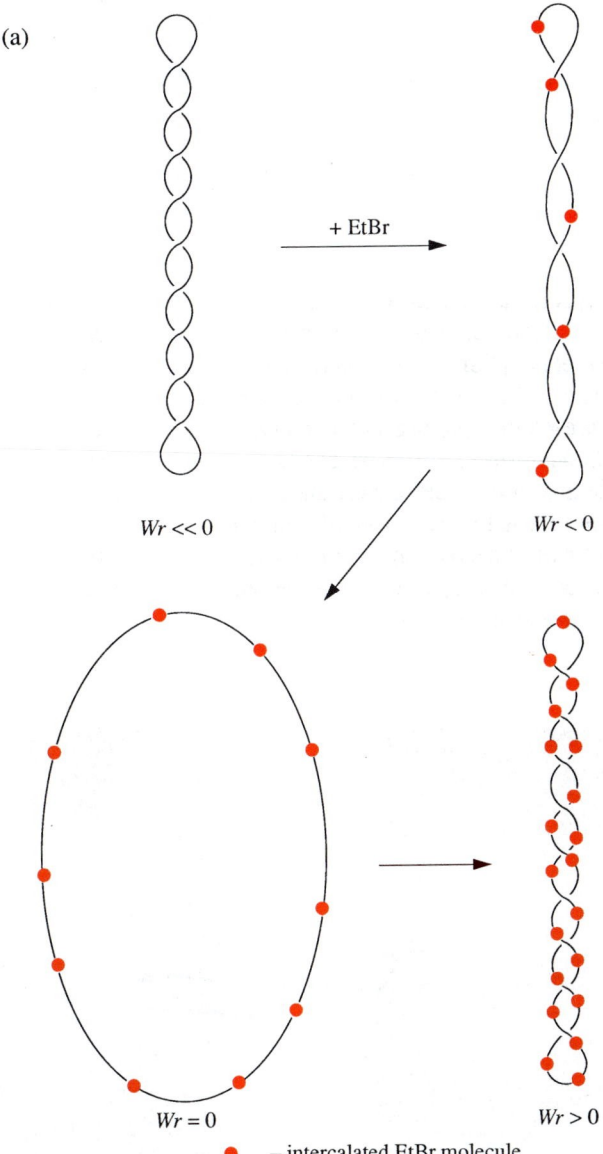

Figure 2.11. The effect of ethidium bromide (EtBr) on closed-circular DNA. **(a)** Intercalation of EtBr into an initially negatively supercoiled DNA ($Wr \ll 0$) causes a reduction in twist, and a concomitant increase in writhe. The alteration in conformation is shown. **(b)** An initially relaxed DNA bound with EtBr becomes positively writhed ($Wr > 0$). Treatment with a topoisomerase removes the positive writhe, and subsequent removal of the EtBr leads to negative writhe (i.e. the DNA becomes negatively supercoiled).

migrates as an unresolved band. Alternatively, if the DNA is initially negatively supercoiled, and hence has negative writhe, the intercalation of EtBr into the double helix will cause a lowering of mobility and the appearance of resolved topoisomers (*Figure 2.12*). At a critical concentration of EtBr, the DNA will appear to be relaxed on the gel; the average writhe of the original molecules being cancelled out by reduced twist caused by intercalation ($Wr = 0$, $Tw = Lk$). With a further increase in EtBr concentration, the DNA will exhibit increasing mobility due to positive writhe in an analogous fashion to the initially relaxed sample described above.

The concentration of EtBr ($\sim$1 μg ml^{-1}) commonly used in gel electrophoresis of, for example, DNA restriction fragments does not allow any resolution of topoisomers of plasmid DNA. Under these conditions, DNA is essentially saturated with EtBr, and all closed-circular molecules have much reduced twist and high, unresolvable, positive writhe. Although nicked-circular DNA binds EtBr in an analogous fashion, the resulting reduction in twist manifests itself merely as a swivelling of one strand around the other, no writhing takes place and the molecule retains the most favourable open-ring structure; its mobility on a gel does not change appreciably. [For practical reasons, the alternative intercalating drug, chloroquine (*Figure 2.10b*), which binds DNA less tightly, is often used in electrophoresis experiments (20).]

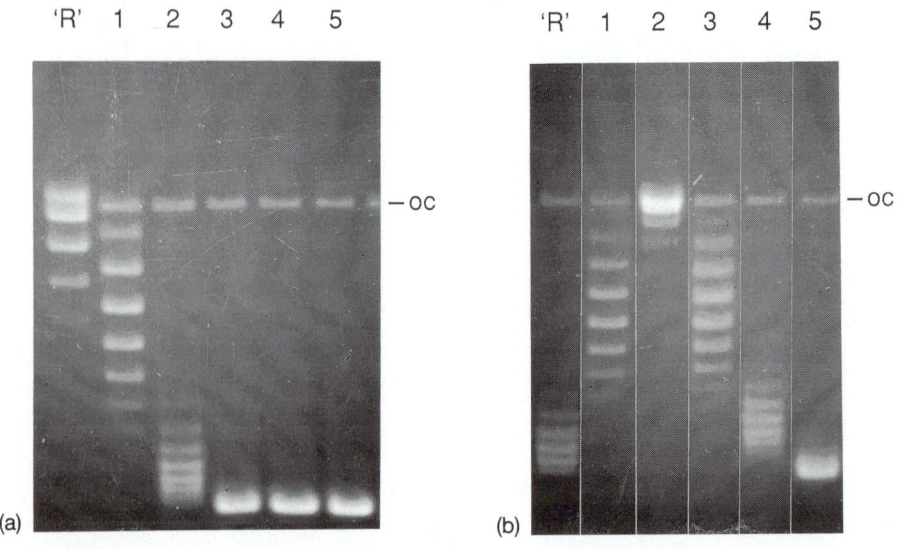

Figure 2.12. Effect of EtBr on gel electrophoresis of closed-circular DNA. **(a)** Agarose gel of samples of pBR322 DNA of increasing negative supercoiling, prepared using the method in *Figure 2.11b*. 'R'; relaxed DNA. 1–5; samples of increasing average negative linking difference, up to $\sigma = -0.07$. **(b)** The same samples, electrophoresed in the presence of EtBr. The relaxed sample ('R') now has positive writhe, and higher mobility. Negatively supercoiled samples (e.g. 3) now have reduced negative writhe and migrate more slowly. Although samples 3–5 appear identical in (a), they are resolved from each other in (b). Sample 1 has the same mobility in (a) and (b), but is negatively writhed in (a) and positively writhed in (b). oc; open-circular DNA.

Another way to think of the effect of intercalators is to consider relaxation of DNA in the presence of EtBr. Consider a topoisomer of 'relaxed' pBR322 DNA ($Lk = Lk_m = +415$, $Tw = +415$, $Wr = 0$); addition of EtBr causes an untwisting of the DNA by, for example, eight turns ($Tw = +407$) (*Figure 2.11b*). The molecule must hence adopt an equivalent positive writhe ($Wr = +8$). Treatment of the sample with a topoisomerase leads to relaxation of the DNA under the changed conditions (added intercalator, reduced twist), that is $Wr \rightarrow 0$; $Lk \rightarrow +407$, $Tw = +407$ (these numbers will not be strictly accurate; see Section 6.1). Subsequent removal of the EtBr causes an increase in twist and the appearance of negative writhe; that is, the molecule becomes negatively super-coiled under the original conditions (e.g. $Lk = +407$, $Wr \approx -6$, $Tw \approx +413$). Such a procedure is commonly used for the preparation of negatively supercoiled DNA of any required average specific linking difference (*Figure 2.12*); relaxation under conditions of increasing EtBr concentration leads to increasing negative supercoiling after removal of the intercalator.

Some other classes of DNA binding molecule have the opposite effect on the DNA helix, that of increasing the twist of the DNA. The best-known example of such a molecule is netropsin [*Figure 2.10c*; (21)], which binds to AT-rich DNA in the minor groove of the double helix, causing an increase in the winding of the helix of around 9° per molecule bound. Netropsin can consequently be used to produce the opposite effect to an intercalator in the experiments described previously, although the relatively low efficiency of the process, and the restricted availability of the material mean it has been much less widely used.

5.4. The effect of protein binding

The binding of certain proteins can also affect the geometry of supercoiled DNA. The best-known example of this process is the winding of DNA around the eukaryotic histone octamer to form the nucleosome. Each nucleosome involves the wrapping of about 146 bp of DNA in 1.8 left-handed turns around the histone core (22). This is a special case of the toroidal winding of DNA described in Section 4.2, and results in the stabilization of negative writhing in the structure of the complex (see Chapter 3, Section 5.1). Hence, an initially relaxed DNA, when incorporated into nucleosomes has negative writhe associated with the protein complex, and must develop compensating positive writhe and/or ΔTw elsewhere in the molecule. Treatment with a topoisomerase will relax these compensatory distortions, with a corresponding negative linking number change. Subsequent removal of the bound protein yields a negatively supercoiled DNA. The average linking difference produced in a relaxed distribution of topoisomers in this sort of experiment is approximately -1 per nucleosome bound. This situation will be considered in more detail in Chapter 3. Other examples of proteins whose binding affects the supercoiling geometry of DNA are DNA gyrase (Chapter 5, Section 3) and RNA polymerase (see Chapter 6, Section 4.1). These should be distinguished from DNA-binding proteins which merely impart a planar bend to the DNA, although it may be that small amounts of twisting or writhing are involved in many DNA–protein interactions. The bending of DNA in response to protein binding has been reviewed by Travers (23).

6. Thermodynamics of DNA supercoiling

The increase in free energy associated with the process of supercoiling has been referred to several times in the preceding sections. The extent of supercoiling (ΔLk) is partitioned into elastic deformations of the axis of the DNA helix (writhe) and of the winding of the double helix itself (twist). The free energy associated with supercoiling is the sum of the contributions of the energies required for these bending and twisting motions. The following sections consider the experiments that have been used to analyse the thermodynamics of this process.

6.1. Ethidium bromide titration

Since the intercalation of EtBr causes changes in the conformation of closed-circular DNA, namely a reduction in twist and a concomitant increase in writhe, and such changes have an energetic cost, the intrinsic binding affinity of EtBr to the DNA double helix is modulated by the topological state of the DNA. In particular, EtBr will bind with higher affinity to a negatively supercoiled molecule ($\Delta Lk < 0$) than to an unconstrained molecule (e.g. nicked-circular), since binding results in a reduction in the energetically unfavourable (negative) writhing of the molecule, as well as the alteration in twist which occurs in both closed-circular and nicked molecules. In the same way, binding of EtBr to relaxed or positively supercoiled DNA is less favourable than to nicked-circular, since the result is the introduction of (positive) writhing. This is the reason that the numbers used in the example in Section 5.4 are not strictly accurate. Relaxation of the DNA in the presence of EtBr causes an altered affinity for the intercalator; the twist after the relaxation will be a little less than before.

This differential binding of EtBr to negatively supercoiled and nicked-circular DNAs was the basis of the original determinations of the free energy associated with the supercoiling process. In the 1970s, the binding isotherms were determined for EtBr binding to both negatively supercoiled DNA and nicked-circular DNA formed from it by DNase I digestion. Bauer and Vinograd (15) measured the alteration of the buoyancy of DNA in caesium chloride gradients caused by titration of EtBr; Hsieh and Wang (24) followed the binding spectroscopically under similar solution conditions. At a critical stoichiometry of bound dye, all the writhing of the closed-circular molecule will be converted to untwisting caused by intercalation (see *Figure 2.11b*), and the closed-circular and nicked molecules will have the same conformation and the same affinity for further binding of EtBr. Below this critical value, a higher total EtBr concentration will be required to achieve a given bound stoichiometry for the nicked molecule, when compared with the closed-circular molecule, and this difference can be used to determine the free energy associated with the initial ΔLk. The critical stoichiometry is directly related to the initial linking difference of the supercoiled molecule by the unwinding angle of EtBr (26°/molecule bound). These experiments showed that the free energy of negative supercoiling in DNA molecules of several thousand base pairs in length had a quadratic dependence on the linking difference, that is, the free energy was approximately proportional to the square of the extent of supercoiling:

$$\Delta G_{sc} = K.\Delta Lk^2 \qquad (2.8)$$

Such a quadratic dependence of ΔG on the extent of deformation of the molecule is expected for a stretching or elastic process (cf. Hooke's Law), and thus suggests that the deformation of the DNA helix structure caused by the supercoiling of naturally occurring DNA is in the elastic range. [Bauer and Vinograd (15) did propose a small cubic term in Equation 2.8, which may be significant at higher levels of supercoiling, but this is not usually considered.]

6.2. The Gaussian distribution of topoisomers

The first use of EtBr titration for the determination of the free energy of supercoiling pre-dated the resolution of individual topoisomers on agarose gels, which was first demonstrated by Keller and Wendel in 1974 (16,17). The realization that the individual bands on a gel represent adjacent topoisomers, and that following ligation of the nicked-circular form (or topoisomerase-mediated relaxation), the relative concentrations of the topoisomers represent an equilibrium distribution, suggested another method for the determination of the free energies.

In 1975 Depew and Wang (5), and Pulleyblank *et al.* (14) resolved distributions of topoisomers of DNAs from 2200 to 9850 bp formed both by ligation of nicked-circular DNA and topoisomerase-mediated relaxation on agarose gels (*Figure 2.13a*), and determined the relative concentrations of the topoisomers from the intensities of the bands (*Figure 2.13b*). The analysis described here is based on that of Depew and Wang (5), although the nomenclature has been updated to that used throughout this book.

For a relaxed distribution of topoisomers prepared by ligation of nicked-circular DNA under given temperature and solution conditions (*Figure 2.13*), the most intense topoisomer band contains DNA of $Lk = Lk_m$. However, as pointed out previously, the formation of this most probable topoisomer usually needs a small (energy requiring) twist and/or writhe displacement from the average conformation of the nicked-circular DNA. This hypothetically most stable conformation corresponds to the average linking number, Lk°, or the centre of the distribution. The small angular displacement required (ω) is hence $Lk_m - Lk^\circ$ ($-0.5 \leqslant \omega \leqslant 0.5$). Assuming that the free energy associated with this displacement is proportional to the square of the displacement (Hooke's Law), the free energy of supercoiling of the Lk_m isomer is:

$$G(Lk_m) = K\omega^2 \qquad (2.9)$$

where K is a constant under the conditions of the experiment. Similarly, for the formation of another species in the distribution, $Lk(x)$, with linking difference (ΔLk) here equal to the integral value $Lk(x) - Lk_m$, the free energy is:

$$G[Lk(x)] = K(\Delta Lk + \omega)^2 \qquad (2.10)$$

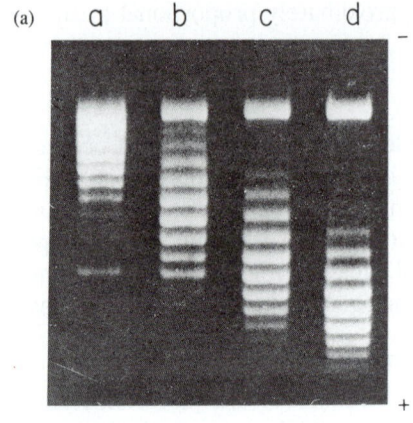

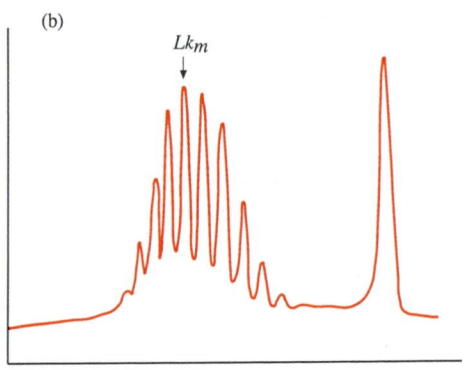

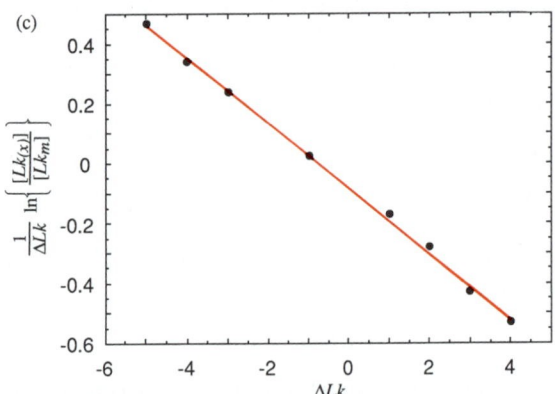

Figure 2.13. The Gaussian distribution of topoisomers. **(a)** Plasmid DNA relaxed at different temperatures. **(b)** Quantitation of the intensities of the bands corresponding to individual topoisomers in (a) 14 °C. **(c)** The data from (b) plotted according to Equation 2.12, to yield K and ω (a reproduced and b, c redrawn from ref. 5).

The linking difference in this case is determined by counting from the most intense band in the distribution ($=Lk_m$). The sign of ΔLk may be determined from the dependence of the mobility of the DNA on different ligation temperatures (5). The standard thermodynamic equation,

$$\Delta G = -RT \, lnK'$$

(the equilibrium constant is written here as K' to distinguish it from the elastic constant K in this derivation) relates the free energy to the concentrations of the equilibrated species, $[Lk_m]$ and $[Lk(x)]$. Hence:

$$G(Lk(x)) - G(Lk_m) = -RT \, ln \frac{[Lk(x)]}{[Lk_m]}$$

$$= K[(\Delta Lk + \omega)^2 - \omega^2] \qquad (2.11)$$

which can be rearranged to yield

$$-\frac{RT}{\Delta Lk} ln \frac{[Lk(x)]}{[Lk_m]} = K(\Delta Lk + 2\omega) \qquad (2.12)$$

The data obtained from densitometer scans of topoisomer distributions such as that illustrated in *Figure 2.13b* may be plotted according to Equation 2.12 to yield a good straight line (*Figure 2.13c*). The concentration distribution of relaxed topoisomers is thus a Gaussian distribution centred around $Lk_m - \omega$ ($=Lk°$). This confirms that the supercoiling of DNA is an elastic process, at least for low values of ΔLk. The result is often extrapolated to higher levels of supercoiling; if ΔLk is large, ω becomes insignificant ($|\omega| \leq 0.5$), and therefore from Equation 2.11:

$$\Delta G_{sc} = K.\Delta Lk^2 \qquad (2.13)$$

In other words, at higher levels of supercoiling, the distinction between $\Delta Lk = Lk - Lk_m$ and $\Delta Lk = Lk - Lk°$ becomes insignificant.

6.3. The effect of DNA circle size

The energy required to introduce a given linking difference into a DNA circle should be inversely proportional to the size of the circle (see Section 3.2). As described, the extent of supercoiling may be normalized to the size of the circle (Section 3.2), and the same applies to the free energy. Hence, dividing by N (number of base pairs) to give the free energy of supercoiling per base pair gives:

$$\frac{\Delta G_{sc}}{N} = NK \left(\frac{\Delta Lk}{N}\right)^2 \qquad (2.14)$$

The constant NK was shown to be largely independent of the size of the DNA circle in the size range investigated. The value determined was approximately $1100 \, RT$. The value of $\Delta Lk/N$ is proportional to the specific linking difference; hence the free energy of supercoiling per base pair is proportional to the square of the specific linking difference, independent of circle size.

6.4. *Supercoiling in small DNA circles*

In the early 1980s, studies of the ligation reactions of linear DNAs yielded interesting results in the case of small DNA fragments, below around 1000 bp. Ligation of linear DNA fragments into circles becomes an increasingly slow process as the DNA size is reduced, reflecting the increasing difficulty of bending a smaller DNA fragment into a circle. Furthermore, below about 500 bp, the rate of ligation becomes a periodic function of the DNA length with a periodicity of around 10 bp (25,26). This implies that the energy required to align the ends of the DNA strands by twisting or writhing, if the length of the DNA is not a multiple of the helical repeat, becomes a significant proportion of the activation energy of the ligation process in small circles. This makes sense because the required change in twist or writhe of up to 0.5 must be distributed over a shorter length of DNA. A corollary of this is that, below 500 bp, the energy required to change Lk by one unit becomes large, and only one, or at most two, topoisomers are visible in the relaxed distribution (e.g. *Figure 2.14*). In the case where two isomers are visible, each isomer has a significant level of positive or negative supercoiling. Hence, the specific linking difference must certainly be related to the centre of the topoisomer distribution, namely $Lk°$, rather than the most intense topoisomer in the distribution (Lk_m), since DNA of $Lk = Lk_m$ may be significantly supercoiled. The specific linking difference (σ) of a topoisomer of a small DNA circle of linking number Lk is hence:

$$\sigma = \frac{Lk - Lk°}{Lk°} \tag{2.15}$$

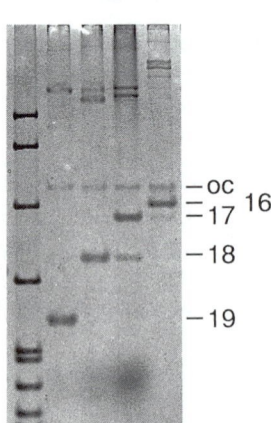

Figure 2.14. Topoisomers of small DNA circles. Tracks 1–4 contain a 196 bp closed-circular DNA, with increasing negative linking difference (the gel is run in the presence of EtBr; all the topoisomers have positive writhe; see *Figure 2.12*). Only one or two topoisomers are visible in each sample. The nicked circular band (oc) is present in all samples. The actual values of Lk for the topoisomers are indicated.

Shore and Baldwin (27) and Horowitz and Wang (28) determined the free energies of supercoiling in small circles (down to a size of 210 bp), from the ratios of relaxed topoisomers using modifications of the procedure outlined in Section 6.2. It was shown that the constant NK (Equation 2.14) increases gradually with decreasing DNA length (*Figure 2.15*). This was interpreted as being due to the increasing unfavourability of writhing relative to twisting of the DNA. A given twisting or untwisting of the helix should require the same energy per unit length of the DNA in any size of circle. The same is not true of writhing, however. Writhing is essentially a bending motion of the DNA helix, and a small DNA circle, even when it lies in a plane, already has significant bending. The further bending caused by writhing of the helix axis requires more energy in a small circle compared with a large one. It is thought, therefore, that supercoiling of small DNA circles is partitioned largely into twisting, rather than writhing. This effect may be modelled using rubber tubing (as always). A very short length of tubing closed into a circle will show, firstly, a much higher resistance to the introduction of a supercoil, and secondly, a higher proportion of twisting than was seen previously (Section 4.2). It has been suggested that moderate linking differences in small DNA circles are partitioned almost exclusively into twisting of the DNA helix. Although this effect undoubtedly plays a part, the more recent results of Bates and Maxwell (29) concerning the coupling of the energy of ATP hydrolysis to DNA supercoiling by DNA gyrase (see Chapter 5, Section 3)

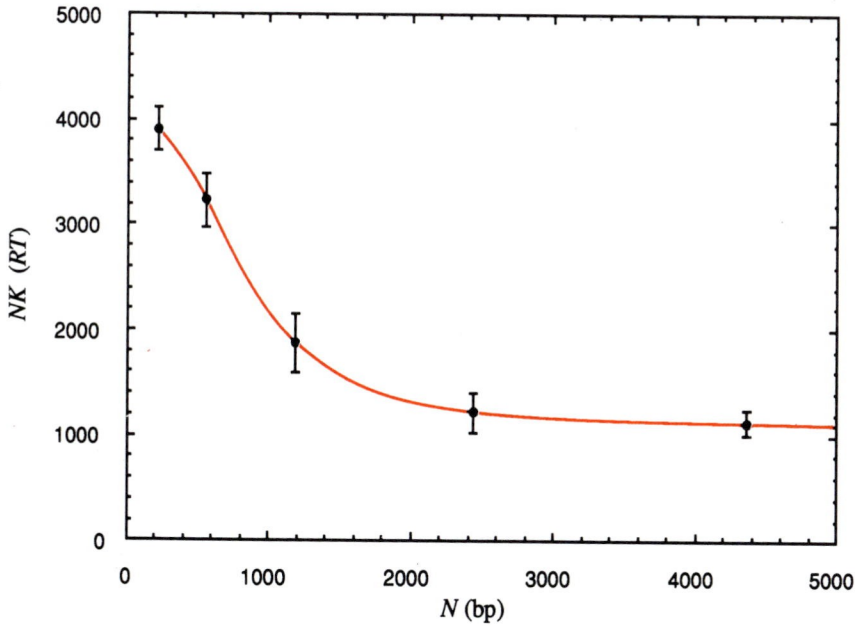

Figure 2.15. Length dependence of the constant NK. The value of NK (Equation 2.14) is plotted against N for circles from 200 to 4500 bp. The bars show estimated errors (redrawn from ref. 28).

suggest that the free energy for the supercoiling of small DNA circles may not be so large as previously determined.

7. Biological effects of supercoiling free energy

The excess free energy associated with the negative supercoiling of DNA may be utilized in many cellular mechanisms. In general, processes which require an untwisting or a writhing of DNA, or which stabilize such deformations, are facilitated with negatively supercoiled, as compared to relaxed, DNA.

Examples of such processes include the replication and transcription of DNA, which require the unwinding of the DNA helix, the formation of nucleosomes, and other protein complexes on DNA, which stabilize the negative writhing of the helix, and the formation of altered DNA structures, such as Z-DNA and cruciforms (see Chapter 1, Sections 2.4 and 3.1). These and other effects of DNA supercoiling and its associated free energy, and their biological significance, will be considered in detail in Chapter 6.

7. Conclusions

This chapter contains the 'basics' for an understanding of DNA supercoiling; the fundamental topological concepts of closed-circularity and linking number, and the geometrical description in terms of twist and writhe. The practical aspects of the behaviour of supercoiled DNA during electrophoresis are important for the interpretation of many experiments where the topology of the DNA is a factor. The excess free energy associated with supercoiling has an important influence on many DNA-associated processes *in vivo*. The final chapter will consider the diverse effects of supercoiling in living cells.

8. Further reading

Bauer,W.R., Crick,F.H.C., and White,J.H. (1980) Supercoiled DNA. *Sci. Am.*, **243 (July)**, 100.

Cozzarelli,N.R., Boles,T.C., and White,J.H. (1990) Primer on the topology and geometry of DNA supercoiling. In Cozzarelli,N.R. and Wang,J.C. (ed.) *DNA topology and its biological effects*. Cold Spring Harbor Laboratory Press, Cold Spring Harbor, p. 139.

Depew,R.E. and Wang,J.C. (1975) Conformational fluctuations of DNA helix. *Proc. Natl. Acad. Sci. USA*, **72**, 4275.

9. References

1. Dulbecco,R. and Vogt,M. (1963) *Proc. Natl. Acad. Sci. USA*, **50**, 236.
2. Weil,R. and Vinograd,J. (1963) *Proc. Natl. Acad. Sci. USA*, **50**, 730.

3. Vinograd,J., Lebowitz,J., Radloff,R., Watson,R., and Laipis,P. (1965) *Proc. Natl. Acad. Sci. USA*, **53**, 1104.
4. Bauer,W.R. (1978) *Annu. Rev. Biophys. Bioeng.*, **7**, 287.
5. Depew,R.E. and Wang,J.C. (1975) *Proc. Natl. Acad. Sci. USA*, **72**, 4275.
6. White,J.H. (1969) *Am. J. Math.*, **91**, 693.
7. Fuller,F.B. (1971) *Proc. Natl. Acad. Sci. USA*, **68**, 815.
8. Fuller,F.B. (1978) *Proc. Natl. Acad. Sci, USA*, **75**, 3557.
9. Crick,F.H.C. (1976) *Proc. Natl. Acad. Sci. USA*, **73**, 2639.
10. Bauer,W.R., Crick,F.H.C., and White,J.H. (1980) *Sci. Am.*, **243 (July)**, 100.
11. Cozzarelli,N.R., Boles,T.C., and White,J.H. (1990) In Cozzarelli,N.R. and Wang,J.C. (ed.) *DNA topology and its biological effects*. Cold Spring Harbor Laboratory Press, Cold Spring Harbor, p. 139.
12. Boles,T.C., White,J.H., and Cozzarelli,N.R. (1990) *J. Mol. Biol.*, **213**, 931.
13. Adrian,M., ten Heggeler-Bordier,B., Wahli,W., Stasiak,A.Z., Stasiak,A., and Dubochet,J. (1990) *EMBO J.*, **9**, 4551.
14. Pulleyblank,D.E., Shure,M., Tang,D., Vinograd,J., and Vosberg,H.-P. (1975) *Proc. Natl. Acad. Sci. USA*, **72**, 4280.
15. Bauer,W. and Vinograd,J. (1970) *J. Mol. Biol.*, **47**, 419.
16. Keller,W. and Wendel,I. (1974) *Cold Spring Harbor Symp. Quant. Biol.*, **39**, 199.
17. Keller,W. (1975) *Proc. Natl. Acad. Sci. USA*, **72**, 4876.
18. Wang,J.C. (1974) *J. Mol. Biol.*, **89**, 783.
19. Pulleyblank,D.E. and Morgan,A.R. (1975) *J. Mol. Biol.*, **91**, 1.
20. Shure,M., Pulleyblank,D.E., and Vinograd,J.K. (1977) *Nucleic Acids Res.*, **4**, 1183.
21. Snounou,G. and Malcolm,A.D.B. (1983) *J. Mol. Biol.*, **167**, 211.
22. Richmond,T.J., Finch,J.T., Rushton,B., Rhodes,D., and Klug,A. (1984) *Nature*, **311**, 532.
23. Travers,A.A. (1989) *Annu. Rev. Biochem.*, **58**, 427.
24. Hsieh,T.-S. and Wang,J.C. (1975) *Biochemistry*, **14**, 527.
25. Shore,D., Langowski,J., and Baldwin,R.L. (1981) *Proc. Natl. Acad. Sci. USA*, **78**, 4833.
26. Shore,D. and Baldwin,R.L. (1983) *J. Mol. Biol.*, **170**, 957.
27. Shore,D. and Baldwin,R.L. (1983) *J. Mol. Biol.*, **170**, 983.
28. Horowitz,D.S. and Wang,J.C. (1984) *J. Mol. Biol.*, **173**, 75.
29. Bates,A.D and Maxwell,A. (1989) *EMBO J.*, **8**, 1861.

3

DNA on surfaces

1. Introduction

As if the complexities of twist and writhe were not enough to deal with, an alternative description of the geometrical aspects of supercoiling has been developed by White and co-workers, which considers DNA to be wrapped upon a real or imaginary surface (1). This formulation, and other discussions of the same problem (2), have caused some confusion, and it must be emphasized that they are complementary to, rather than being an improvement on, the normal twist and writhe approach. The development of this method of analysis grew from the realization that in a number of important structures DNA lies on a protein surface. The clearest example of this is the nucleosome (see Chapter 2, Section 5.4) but others include DNA gyrase (Chapter 5, Section 3) and *Escherichia coli* RNA polymerase (3). In addition, in some conformations of DNA alone, the molecule can be considered to lie on an imaginary or virtual surface.

Twist and writhe, although rigorously describing the geometry of supercoiled DNA, are not easy to measure. Indeed, for all but a few simplified cases (4), the calculation of the exact values of twist and writhe involves prohibitive numerical calculations. The surface linking model allows the linking number of DNA to be partitioned into two integer values, the winding number and the surface linking number, which describe, respectively, the winding of the DNA double helix relative to the surface on which it lies, and the topology of that surface in space. This chapter discusses the details of this geometric description of DNA, and some of the confusions it can cause, as well as its application to particular problems.

2. The helical repeat of DNA

The key to the discussion of the surface linking model lies in the concept of the helical repeat of DNA, h, a quantity which we have already discussed in the first two chapters. The easiest way to envisage the measurement of the helical repeat of linear DNA is to imagine the helix lying on a plane surface. h is then the average periodicity of the appearance of one of the DNA strands on the upper face of the helix, away from the surface beneath, measured in base pairs. In practice, h can be measured in just this way, by utilizing DNA adsorbed to a perfectly flat calcium phosphate surface, and allowing the backbone to be cleaved by DNase I, which cuts more effectively at phosphodiester bonds away from the surface [*Figure 3.1*; (5)]. This is an application of the general method of DNA footprinting, in which material (e.g. protein or drugs) bound to DNA protects it from

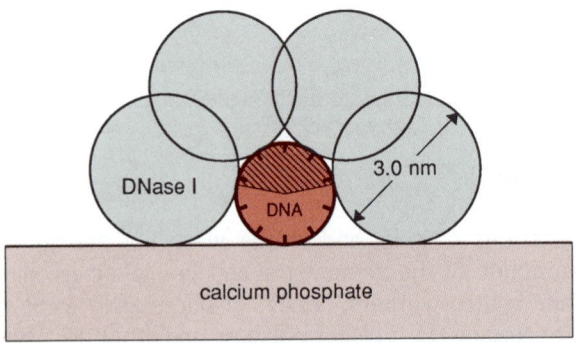

Figure 3.1. Measurement of the helical repeat of DNA by DNase I digestion. DNA (small circle) is shown bound to calcium phosphate (shaded area). The sites of possible cleavage of the DNA by DNase I (large circles) are indicated by the hatched area (redrawn by permission from ref. 5. Copyright © 1980 Macmillan Magazines Limited).

digestion by an enzymic or chemical agent at particular sites (6). Thus the average periodicity of cutting of one of the DNA strands, as measured by the length distribution of the products, is a direct measure of the helical repeat. As seen in Chapter 2, Section 4.2, the average helical repeat is related to the twist of the DNA:

$$Tw = \frac{N}{h} \tag{3.1}$$

This is logical in that the helical repeat and the twist are measures of the winding of the DNA strands about the helix axis. However, most DNA, when supercoiled or lying on a protein surface, for example in the nucleosome, does not lie in a plane. Is it possible to measure the helical repeat of the DNA in such a situation? If the twist of the DNA is known, then one method would be to assign the average helical repeat according to Equation 3.1. However, as mentioned, calculation of the twist is not an easy matter. Twist is defined in terms of the instantaneous rotation of a point on one DNA strand around the tangent to the helix axis at that point, integrated over the whole length of the DNA (7). Since the effect of writhing the DNA is to change the tangent to the helix axis continuously, this complicates the calculation [this is in addition to the requirement that $Tw + Wr = Lk$ if the DNA is a closed-circle (4)].

There is also a problem in the general measurement of h by the DNase I cleavage method, since the cleavage must be determined relative to a surface on which the DNA lies (*Figure 3.1*). In other words, by choosing a surface relative to which the periodicity of nuclease cleavage is measured, such as a curved protein surface, a frame of reference for the measurement is implicitly defined. The helical repeat measured by this method depends on the surface chosen, in addition to any physical twisting or untwisting of the DNA helix.

The helical repeat of the DNA given by DNase I cleavage periodicity need not be the same as that derived from the twist of the DNA (i.e. N/Tw), and may or may not reflect an absolute physical twisting or untwisting of the DNA helix. The winding of DNA around, for example, the nucleosome can thus be described in principle using two frames of reference: by the twist of the DNA, which is measured relative to the local helix axis at any point (7) and by the helical repeat of the DNA relative to the surface on which it lies (i.e. in the surface frame) (1).

It seems worth defining h_t for the twist-related, and h_s for the surface-related helical repeats. Whilst h_s has the virtue that it is in principle an experimentally measurable quantity, h_t has the advantage that it better reflects the actual local twisting up or untwisting of the DNA helix in response to superhelical stress, or protein binding, or whatever. When DNA lies in a plane, $h_s = h_t$, but this need no longer be true if the helix axis is out of the plane. The value of h_t is a property only of the DNA conformation, whereas h_s is also a function of the chosen surface.

It has been suggested by thermodynamic calculations that the bending of the DNA axis caused by writhing can affect the twist of the DNA due to a physical deformation of the structure (8). That is, the act of bending may lead to a stable structure with a reduced helical repeat (increased twist). This effect, if present, may or may not be comparable in magnitude to the apparent change in helical repeat caused by the choice of surface, or frame of reference. However, the binding of DNA to a protein can certainly affect all the parameters describing DNA conformation, as can be seen in the case of the nucleosome (Section 5).

3. The surface linking treatment

The surface linking treatment of DNA supercoiling, as its name implies, uses the surface frame of reference, since this has the virtue that in some cases, most notably in the nucleosome, the surface related helical repeat can actually be measured, and the geometry of the surface on which the DNA lies can in principle be determined. The following discussion is based on White *et al.* (1).

3.1. DNA winding number (Φ)

The winding number of a closed-circular DNA, Φ, describes the winding of the DNA helix around its axis in the surface frame of reference. This parameter may be defined by reference to *Figure 3.2*, which describes the geometry of a DNA whose axis, A lies on a general surface, M. The curve C represents one of

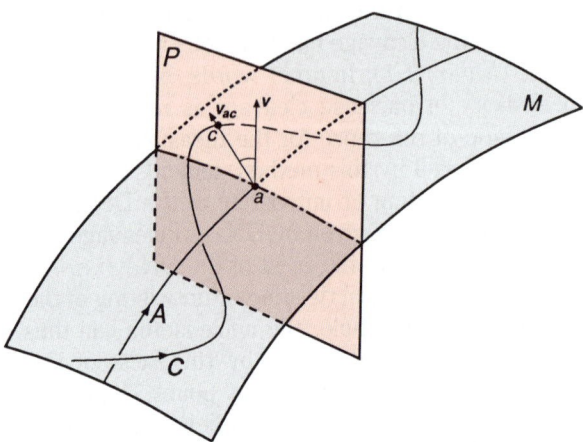

Figure 3.2. Definition of the winding number (Φ). The winding number (Φ) is given by the number of rotations of vector $\mathbf{v}_{ac}$ relative to vector $\mathbf{v}$ as point a traverses the DNA axis, A. See text for further details (redrawn from ref. 1. Copyright 1988 by the AAAS).

the DNA strands, winding in a right-handed helical sense around A, and passing above and below M. At any point, a, along A, a plane, P, is drawn perpendicular to A. P is crossed once by curve C at point c, and the vector $\mathbf{v}_{ac}$ in the plane is the unit vector along the line joining a and c. As P advances along A, $\mathbf{v}_{ac}$ remains perpendicular to A and rotates around it. The extent of rotation is measured relative to the surface normal vector $\mathbf{v}$, which also lies in plane P. For closed-circular DNA, A is a closed curve, and Φ is the number of times $\mathbf{v}_{ac}$ rotates around $\mathbf{v}$ as A is traversed once. Clearly, Φ must be an integer. Φ is positive if the direction of rotation is right-handed, as in *Figure 3.2*. The average helical repeat relative to the surface, h_s, may thus be defined in terms of the DNA winding number:

$$h_s = \frac{N}{\Phi}$$
(3.2)

3.2. Surface linking number (SLk)

If the DNA closed-circle is planar, and the axis A lies in a plane (i.e. M is a plane surface), the reference vector $\mathbf{v}$ will always point upward. The winding number will in this case be equal to Lk, the number of turns of the helix around the axis, (and incidentally to the twist, since $Wr = 0$; *Figure 3.3a*). In general, however, the helix axis A need not lie in a plane, and the reference vector $\mathbf{v}$ need not have a constant direction. In such cases, the linking number will not in general be equal to the winding number, but will also include a contribution from the change in the reference vector, that is, from the topology of the surface M. This contribution, called the surface linking number (SLk), is in simplistic terms, the number of revolutions made by the vector $\mathbf{v}$ in space, as the DNA axis is traversed. SLk is defined as follows: if a new curve A_ε is formed by a small displacement from A

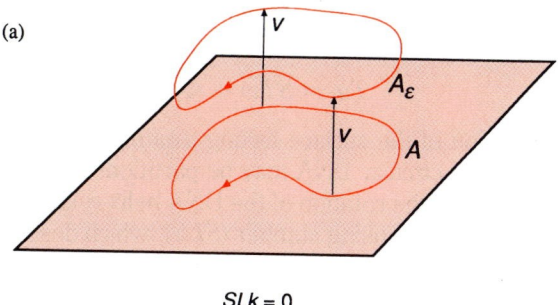

(a)

$SLk = 0$

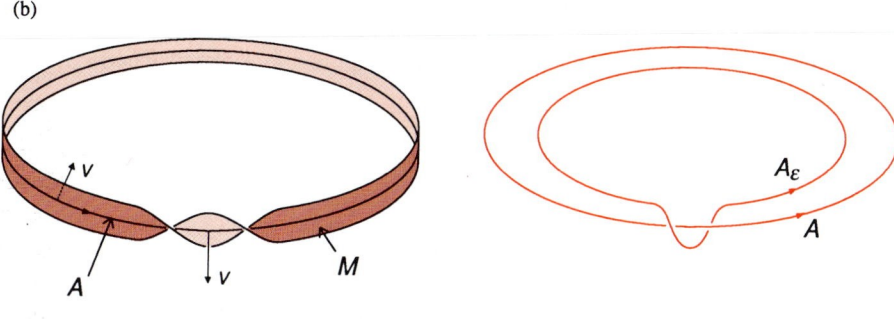

(b)

$SLk = +1$

Figure 3.3. Definition of the surface linking number (SLk). The surface linking number (SLk) is the linking number of the DNA axis A and the curve A_ε, displaced along the surface normal, v. (a) Refers to DNA lying on a plane surface. (b) Refers to DNA lying on a twisted ribbon. See text for further details (redrawn from ref. 1. Copyright 1988 by the AAAS).

along the direction of the normal vector v, SLk is the linking number of the two curves A and A_ε. For example, in *Figure 3.3a*, the DNA axis lies in a plane, the displaced curve A_ε lies entirely above the plane, and $SLk = 0$. In *Figure 3.3b*, the axis curve A lies on a strip surface, M, which is twisted once in a right-handed sense. Hence the surface normal, v makes one right-handed turn around A as the axis is traversed. The displacement curve A_ε and A have a linking number of $+1$ (see Chapter 2, Section 3.1), and hence $SLk = +1$.

The surface linking number is a function of the writhe of the DNA helix axis, since it is in part a property of the trajectory of the axis. The two terms are related by the following equation:

$$SLk = STw + Wr \qquad (3.3)$$

where STw, the surface twist, is the twist of the displaced curve A_ε about the axis curve A. This equation thus has the same form as $Lk = Tw + Wr$ (Chapter 2, Section 4.1). The surface twist, STw is also the correction between the measures of helical winding in the two frames of reference, the twist, Tw and the winding number, Φ.

$$Tw = STw + \Phi \qquad (3.4)$$

The basic formulation of the surface linking treatment is thus that the linking number (Lk) of a closed-circular DNA may be partitioned into the DNA winding number (Φ), a measure of the rotation of the DNA helix relative to the surface on which it lies, and the surface linking number (SLk), which describes the topology of that surface in space; that is:

$$Lk = SLk + \Phi \qquad (3.5)$$

For rigorous proof of the above relationship, see White *et al.* (1). It is also useful to incorporate the specific linking difference (σ) into this formulation. The linking difference of a closed-circular DNA is $Lk - Lk^\circ$ (Chapter 2, Section 3.2), where $Lk^\circ = \Phi^\circ = N/h^\circ$ (for a relaxed DNA whose axis lies in a plane; *Figure 3.3a*). From Equation 3.5,

$$\Delta Lk = Lk - Lk^\circ = SLk + \Phi - Lk^\circ \qquad (3.6)$$

and therefore,

$$\sigma = \frac{\Delta Lk}{Lk^\circ} = \frac{SLk}{Lk^\circ} + \frac{\Phi - Lk^\circ}{Lk^\circ} \qquad (3.7)$$

It is possible now to introduce the surface helical repeat h_s ($=N/\Phi$), and h° ($=N/Lk^\circ$):

$$\sigma = \frac{SLk}{Lk^\circ} + \frac{h^\circ}{h_s} - 1 \qquad (3.8)$$

this equation may be solved for h_s:

$$h_s = \frac{h^\circ}{\sigma - \dfrac{SLk}{Lk^\circ} + 1} \qquad (3.9)$$

In principle, this equation allows the calculation of the surface-related helical repeat for DNA of any specific linking difference, providing the surface linking number of the surface is known.

Since the surface linking number is a topological property of the path followed by the DNA on a surface, it follows that smooth deformation of that surface, that is stretching or bending, without breaking, will not change its value, as long as the DNA remains on the surface during the deformation. Since Lk is also topological invariant, then the winding number and hence the average helical repeat also do not change on smooth deformation. In the following sections, potential applications of this approach will be considered.

As mentioned previously, in a number of situations, DNA either does lie on a surface, for example that formed by a protein, or can be considered to lie on a virtual surface, and hence the surface linking treatment described above can be applied to a number of 'real' problems.

4. Interwound supercoiled DNA

The most stable form adopted by negatively supercoiled DNA in solution is the interwound, or plectonemic form (see Chapter 2, Section 4.2.1). This conformation of DNA may be modelled as if the axis of the DNA helix lies on a cylinder with capped ends, as in *Figure 3.4*. The DNA axis A winds up and down the cylinder, and crosses the hemispherical caps at top and bottom; the actual dimensions of the cylinder are not too important. Firstly, considering the value of the surface linking number for such a structure, the surface normal, v, will always point out of the cylinder, and hence the displaced curve A_ε will always lie

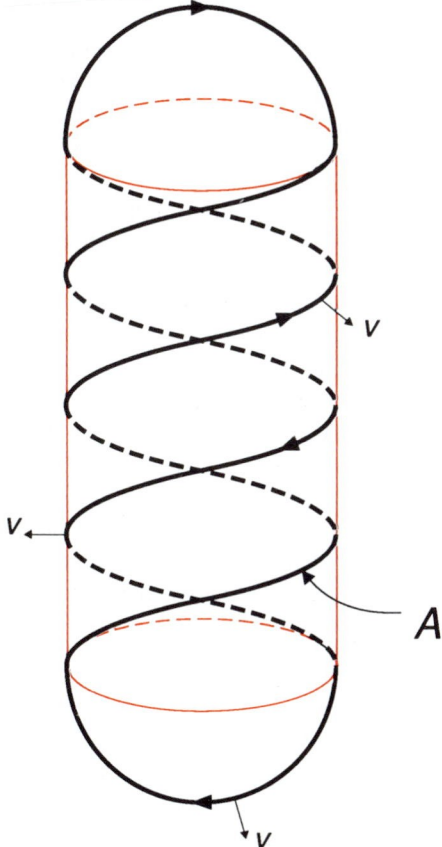

Figure 3.4. Representation of an interwound DNA helix. A, the axis of the DNA double helix, is wound on the surface of a cylinder with capped ends. v is the surface normal vector (see text; redrawn from ref. 1. Copyright 1988 by the AAAS).

outside the cylinder. Clearly, A and A_ε will not be linked, and hence $SLk = 0$. This result means that from Equation 3.5:

$$Lk = \Phi$$

and the surface helical repeat, h_s, from Equation 3.9, simplifies to:

$$h_s = \frac{h°}{\sigma + 1}$$

This means that the apparent helical repeat of the DNA is a function simply of the initial helical repeat and the specific linking difference of the DNA. As an example, σ for naturally occurring plasmid DNA is typically -0.06 (Chapter 2, Section 3.2). If the initial helical repeat of DNA ($h°$) is 10.5 bp/turn, h_s comes out as 11.2 bp/turn. It must be emphasized again that this apparent change is a result of the changed frame of reference of the measurement and does not reflect the physical untwisting of the DNA implied by such a change in helical repeat, although as noted previously, it is likely that the DNA is somewhat untwisted in a plectonemically wound supercoil (Chapter 2, Section 4.2.1). This confusion of a change in helical repeat measured relative to a (virtual) surface, compared with the natural helical repeat as envisaged and measured for a linear or planar length of DNA, is the major misunderstanding which has been caused by the surface linking treatment.

Furthermore, because of the requirement mentioned earlier, that the winding number is unchanged on smooth deformation, any change in the conformation of the DNA which results only in a smooth deformation of the virtual surface, leaves the winding number, and hence average h_s unchanged. Such a deformation would include that caused by changing conditions, such as temperature or ionic strength, or even the addition of intercalating drugs. Even though the effect of intercalators is to untwist the helix of the DNA, the effect would be to alter the winding of the DNA on the cylinder (through a change in writhe; remember, the cylinder is allowed to change shape) so as to leave the apparent helical repeat unchanged. It can therefore be seen that the surface linking treatment of DNA supercoiling does not really produce useful information in the case of interwound DNA, since the winding number is not readily measurable, and is, in any case, not sensitive to the changes in conformation usually seen in this kind of DNA molecule (see Chapter 2, Section 5.3).

5. The nucleosome

As an additional example, the long-standing problem of the conformation of DNA when bound in the nucleosome will be considered (*Figure 3.5*). From the X-ray crystallographic studies of nucleosome core particles by Klug and co-workers (9,10), it is known that 146 bp of DNA wraps itself in about 1.8 turns round the histone octamer core, in a shallow left-handed superhelix. The dimensions of the DNA path are shown in *Figure 3.5*.

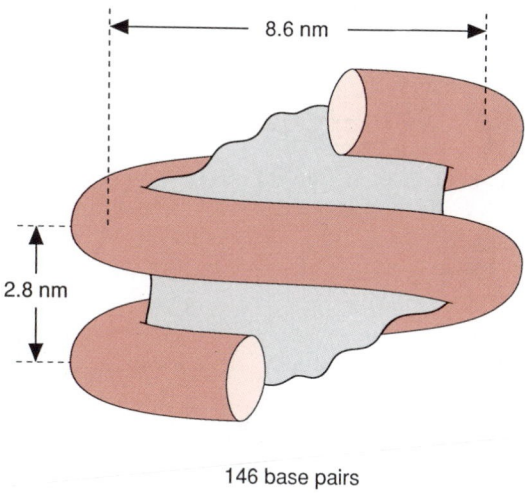

146 base pairs

1.8 superhelical turns

Figure 3.5. A schematic view of the path of DNA around the histone octamer in the nucleosome core particle.

5.1. Linking difference and the 'linking number paradox'

Since DNA is wrapped around the histone octamer in a superhelical path, it should come as no surprise that this writhing of the DNA can contribute to supercoiling if the nucleosomes are part of a closed-circular DNA. Indeed, if a closed-circular DNA is relaxed with a topoisomerase in the presence of a number of nucleosomes (see Chapter 2, Section 5.4), the DNA can be shown, when the histones are removed, to have a linking difference (ΔLk) corresponding to approximately -1 per nucleosome present [the most up-to-date value is -1.01 ± 0.08 (11)]. This means that each nucleosome contributes a linking difference (ΔLk) of approximately -1. Although linking number is undefined for non-closed-circular DNA molecules, it is useful to consider such linking differences for segments of a closed molecule, which must sum to the Lk. Although the ΔLk attributable to one nucleosome is close to 1, there is no requirement for a linking difference in this context to be an integer, and the ΔTw and ΔWr of the length of DNA being considered (relative to the twist and writhe of the equivalent linear unconstrained DNA) will still add up to the ΔLk (Equation 2.4). The same argument also applies to Φ, SLk and ΔLk for a given DNA segment; that is:

$$\Delta Lk = \Delta SLk + \Delta \Phi \qquad (3.10)$$

for a segment of a closed-circular DNA, for example that bound in a nucleosome.

The superhelical turns of DNA around the nucleosome core are of the toroidal type defined in Chapter 2, Section 4.2.1 (*Figure 2.7e*). This can be seen by considering the representation in *Figure 3.6*. A closed-circular DNA incorporated into nucleosomes can be considered to lie wholly on the surface of a torus.

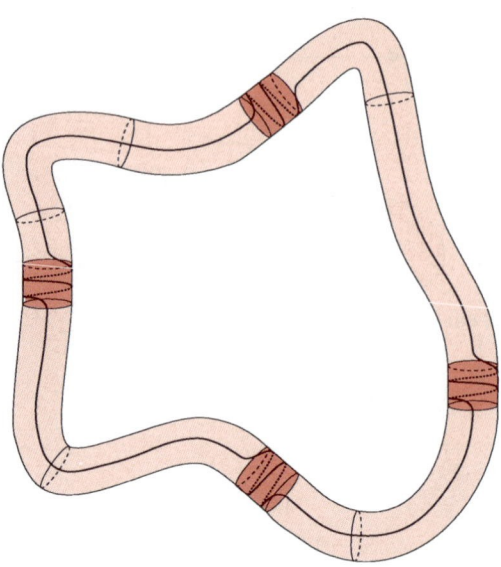

Figure 3.6. Toroidal supercoiling in nucleosome-wrapped DNA. The path of the DNA helix in an array of nucleosomes on closed-circular DNA can be assumed to lie on the surface of an imaginary torus (redrawn from ref. 1. Copyright 1988 by the AAAS).

The wrapping of DNA around the histones corresponds to turns around the torus, and the DNA linking the nucleosomes together lies along the line of the torus. From the definition of the surface linking number (Section 3.2), it can be seen that the total number of turns around the torus equals the SLk for the whole array of nucleosomes. This can be divided into a contribution from each nucleosome, which corresponds to ΔSLk in Equation 3.10. From the crystal structure (10), this value is -1.8.

The so-called 'linking number paradox' refers to the discrepancy between the 1.8 turns of shallow superhelix of DNA around the histone core, which implies a contribution to linking number of roughly -2, and the measured linking contribution, which is close to -1 per nucleosome (9,12). Of course, although a linking difference can be defined as a characteristic of a single nucleosome structure, it is only possible to make a measurement of ΔLk in a closed-circular DNA, ideally one containing several nucleosomes. This means that it is not known for sure whether the linking difference occurs within the structure of the nucleosome (*Figure 3.5*), or whether part of the effect reflects a superhelical ordering of the DNA which joins the nucleosomes together in the closed-circular array (*Figure 3.6*).

5.2. Early resolution of the linking number paradox

When the structure of the nucleosome was first elucidated in 1977 (9), and the nature of the linking number paradox discerned, it was soon reasoned that an alteration in the helical repeat of the DNA in the nucleosome to 10.0 bp/turn from the value in free DNA would resolve the paradox (9,12). This analysis was based on the twist and writhe parameters. The discrepancy between the linking difference on nucleosome formation (~ -1) and the writhe associated with the structure (~ -1.8) was interpreted as a change in the twist of the DNA bound to the histone core. A straightforward calculation yielded the predicted value for the twist-related helical repeat (h_t) of 10.0 bp/turn.

To summarize, one possible explanation for the linking number paradox is that the DNA joining the nucleosomes in a closed-circular array contributes a linking difference of around +0.8/nucleosome, to account for the difference between the contribution from the path of the DNA axis and the measured value. An alternative explanation is that there is a change of helical repeat of the DNA bound to the nucleosome, from 10.5 bp/turn in free DNA to 10.0 bp/turn in nucleosome-wrapped DNA.

5.3. The helical repeat at the nucleosome surface

At about the same time as the above calculations were made, the first accurate values of the periodicity of DNA on the nucleosome were made by DNase I digestion (12–14). Earlier experiments had yielded a value of roughly 10 bp/turn (15). However, in the more accurate determinations, the average value came out as 10.4 bp/turn, not dissimilar to that in free DNA, rather than the predicted 10.0 bp/turn (Section 5.2). However, this result and the prediction were reconciled by consideration of possible steric effects on DNase I digestion of the adjacent turns of helix on the nucleosome (12), due to the large size of the DNase I molecule relative to the DNA helix (see *Figure 3.1*).

As has been seen from consideration of the surface frame of reference, the periodicity of the DNA helix relative to a surface on which it lies corresponds rigorously to h_s and is a function of the winding number, Φ. More recently, estimates of h_s have been made using alternative methods which are not susceptible to the steric problem previously described. These data, and analysis using the winding number and surface linking number, make possible a convincing estimate of the topological and geometric parameters describing the path of DNA around the nucleosome. (Some readers may find this discussion a little complex; if so, they should turn to the summary presented in Section 5.6.)

5.4. Measurements of h_s for the nucleosome

When DNA is wound tightly around the nucleosome core, one of the consequences is the compression of the major and minor grooves (see Chapter 1, Section 2.2) on the inside and the widening of the grooves on the outside of the DNA curve. Certain sequences, notably runs of A–T base pairs, having an intrinsically narrow minor groove, should be most favourably positioned with their minor grooves on the inside of such a sharply curved segment of DNA. Analogously, runs of G–C base pairs should be more favourably aligned with minor grooves facing outwards (16). In 1985, Drew and Travers (17) utilized this hypothesis in a study of the probabilities of the positioning of certain sequences in nucleosome-bound DNA. The periodicity of the appearance of short regions of A–T and C–G nucleotides along many nucleosome-forming sequences were averaged to yield a value of 10.17 ± 0.05 bp. Crucially, the periodicities of A–T and of G–C sequences were exactly out of phase, the A–T sequences appearing with minor grooves on the inside, as predicted. This value of 10.17 bp therefore reflects the periodicity of appearance of these sequences on the inside or outside of the DNA curve, relative to the surface of the histone core. In other words, this is a direct measure of h_s [the value of 10.17 bp/turn was later confirmed by direct sequencing of many nucleosome-forming sequences (18)].

Recently, an alternative DNA digestion technique has been used to determine the helical periodicity of DNA on the nucleosome (19,20). Hydroxyl radical cleavage is not subject to the steric constraints associated with DNase I, and the extent of cleavage is modulated both by the protection of bound protein and by the width of the minor groove of the DNA helix (see above). The variation in such cleavage should thus provide an ideal measure of the periodicity of the DNA helix relative to a protein surface, that is h_s. Measurement of helical periodicities by hydroxyl radical cleavage in a single nucleosome (19) and in mixed sequence nucleosomes (20) have yielded an average value of 10.18 ± 0.05 bp/turn, a value in remarkably close agreement with that derived from sequence periodicities.

5.5. Topology and geometry of DNA in the nucleosome

The parameter measured by the experiments described above, h_s (=10.18 bp/turn), and the dimensions of the nucleosomes in *Figure 3.5*, allow the determination of the winding number Φ of nucleosome-wrapped DNA (Equation 3.2):

$$\Phi = \frac{N}{h_s} = \frac{146}{10.18} = 14.34$$

The winding number of 146 bp of linear DNA is $146/10.49 = 13.92$; the average value of $h°$ was calculated by Hayes *et al.* (19) for the specific fragment studied as 10.49 bp/turn. Therefore, the change in winding number on nucleosome formation is:

$$\Delta\Phi_{nuc} = +0.42$$

The DNA is wrapped in 1.8 left-handed turns around the nucleosome core; hence ΔSLk, the change in surface linking number on nucleosome formation is -1.8 (SLk for planar, i.e. uncomplexed DNA, is zero, *Figure 3.3*). Hence, from Equation 3.10, the predicted linking difference attributable to the formation of one nucleosome can be calculated:

$$\Delta Lk_{nuc} = \Delta SLk + \Delta\Phi = -1.8 + 0.42 = -1.38 \; (\pm 0.13; \text{ ref. } 19)$$

Essentially the same analysis has also been presented by Travers and Klug (2), except that no correction has been made for the difference between the number of superhelical turns around the nucleosome (SLk), and the corresponding writhe of the DNA, which are not identical (Equation 3.3); this leads to a slightly different result. The use of SLk in this calculation (and in Hayes *et al.*) removes this problem. In the terminology used by Travers and Klug, the 'local' frame of reference is the equivalent of the surface frame; this analysis pre-dates the formal description of surface linking number, and should probably be recast in surface linking terms.

When the calculated value for ΔLk on nucleosome formation (-1.38 ± 0.13) is compared with the experimentally determined value of -1.01 ± 0.08, it can be seen that this analysis does not resolve the linking number paradox. To account for this discrepancy, it may be necessary to return to the suggestion that the DNA between successive nucleosomes in closed-circular DNA is organized so as to provide a contribution to the ΔLk (Section 5.1). Hayes $et\ al.$ (19) point out that more than 146 bp of DNA may be organized by the histone core, on the basis of their hydroxyl radical experiments. If this effect accounts for all of the discrepancy, then each linker DNA in a nucleosome array contributes $+0.37$ to the linking number, a considerably smaller value than that required in the original analysis (Section 5.2). Alternatively, it has been suggested by White $et\ al.$ (21,22) that small distortions in the shape of nucleosomes may lead to significant alterations in the surface linking number (SLk) attributable to the nucleosome structure.

Although these findings have not completely resolved the linking number paradox, they have shown that the intrinsic periodicity of DNA does change on binding to the nucleosome core. As previously stated, the winding number, Φ, is not the same as the twist of the DNA (Section 3.2); they are measured in different frames of references, and are related by the surface twist, STw (Equation 3.4). The equation relating the change in these quantities on nucleosome formation is:

$$\Delta Tw = \Delta STw + \Delta\Phi \qquad (3.11)$$

ΔSTw is a function of the geometry of the nucleosome ($Figure\ 3.5$) and is calculated to be -0.19 (4,23). $\Delta\Phi = +0.42$ (see above); hence:

$$\Delta Tw_{nuc} = -0.19 + 0.42 = +0.23$$

In other words, the DNA is physically twisted up on nucleosome formation. In terms of the twist-related helical repeat, h_t, (Section 2) this corresponds to a helical repeat of 10.32 bp/turn in the frame of reference of the DNA axis. There are two possible physical explanations for this over-twisting of the helix. As already mentioned (Section 2), thermodynamic calculations have suggested that bending of the DNA helix axis can intrinsically stabilize a conformation with an increased twist (reduced helical repeat) (8). On the other hand, other work has implied that the interaction with the histone proteins of the nucleosome core may force the DNA into an over-twisted conformation (24). Hayes $et\ al.$ prepared nucleosomes on a DNA fragment containing rigid oligo(dA)·oligo(dT), flexible oligo[(A–T)], and intrinsically curved segments (see Chapter 1, Section 4), which have distinctly different structural parameters, including helical repeats. They found that the different segments adopted a very uniform helical repeat, h_s, when bound in nucleosomes, as measured by hydroxyl radical cleavage. This implies that the interaction with the histone core, rather than any intrinsic property of the DNA may be the major determinant of the helical repeat on the nucleosome surface.

5.6. Summary

The accumulating data on the geometry of the nucleosome and measurements of the periodicity of DNA on the nucleosome, when analysed using the surface linking method have led to the following conclusions. The change in linking number directly attributable to the formation of the nucleosome core on a closed-circular DNA is calculated to be -1.38 [based on (19)]. The experimentally determined value for an array of nucleosomes is -1.01 (11). This discrepancy may be accounted for either by superhelical ordering of the DNA linking the nucleosomes in the array, or, possibly, by distortions in the shape of the nucleosome (21,22). These results imply that the DNA is physically twisted up on nucleosome formation, by $+0.23$ turns over the whole DNA wrap; this corresponds to a helical repeat in the frame of the helix axis (given by N/Tw) of 10.32 bp/turn, compared with 10.49 bp/turn, for the same fragment studied, when uncomplexed. The twisting up of the helix may be a direct result of the required bending of the helix on nucleosome formation (8), or may be a response to specific interactions with the histone octamer core (24).

6. Conclusions

This chapter has described an alternative geometrical description of the super-coiling of DNA, which is complementary to the twist and writhe formulation described in Chapter 2. Whilst not very useful for the analysis of supercoiling of free DNA circles, it can be very helpful in discussing the complexities of DNA helically wound on a protein surface. An alternative application, not discussed here, is to DNA catenanes (see Chapter 4) toroidally wound about each other (25). The biggest difficulty with the surface linking approach is the confusion resulting from the seductive, but false, equation of the winding number, Φ, and the twist Tw of a closed-circular DNA, or referral to the 'helical repeat' of the DNA, without being specific about the frame of reference used.

7. Further reading

Cozzarelli,N.R., Boles,T.C., and White,J.H. (1990) In Cozzarelli,N.R. and Wang,J.C. (ed.) *DNA topology and its biological effects*. Cold Spring Harbor Laboratory Press, Cold Spring Harbor, p. 139.

White,J.H., Cozzarelli,N.R., and Bauer,W.R. (1988) *Science*, **241**, 323.

White,J.H., Gallo,R.M., and Bauer,W.R. (1992) Closed-circular DNA as a probe for protein-induced structural changes. *Trends Biochem. Sci.*, **17**, 7.

8. References

1. White,J.H., Cozzarelli,N.R., and Bauer,W.R. (1988) *Science*, **241**, 323.
2. Travers,A.A. and Klug,A. (1987) *Phil. Trans. R. Soc. Lond.*, **B317**, 537.
3. Wang,J.C., Jacobsen,J.H., and Saucier,J.M. (1977) *Nucleic Acids Res.*, **4**, 1225.
4. White,J.H. and Bauer,W.R. (1986) *J. Mol. Biol.*, **189**, 329.
5. Rhodes,D. and Klug,A. (1980) *Nature*, **286**, 573.
6. Tullius,T.D. (1989) *Annu. Rev. Biophys. Biophys. Chem.*, **18**, 213.
7. Fuller,F.B. (1978) *Proc. Natl. Acad. Sci. USA*, **75**, 3557.
8. Levitt,M. (1978) *Proc. Natl. Acad. Sci. USA*, **75**, 640.
9. Finch,J.T., Lutter,L.T., Rhodes,D., Brown,R.S., Rushton,B., Levitt,M., and Klug,A. (1977) *Nature*, **269**, 29.
10. Richmond,T.J., Finch,J.T., Rushton,B., Rhodes,D., and Klug,A. (1984) *Nature*, **311**, 532.
11. Simpson,R.T., Thoma,F., and Brubaker,J.M. (1985) *Cell*, **42**, 799.
12. Klug,A. and Lutter,L.C. (1981) *Nucleic Acids Res.*, **9**, 4267.
13. Lutter,L.C. (1979) *Nucleic Acids Res.*, **6**, 41.
14. Prunell,A., Kornberg,R.D., Lutter,L., Klug,A., Levitt,M., and Crick,F.H.C. (1979) *Science*, **204**, 855.
15. Noll,M. (1974) *Nucleic Acids Res.*, **1**, 1573.
16. Drew,H.R. and Travers,A.A. (1984) *Cell*, **37**, 491.
17. Drew,H.R. and Travers,A.A. (1985) *J. Mol. Biol.*, **186**, 773.
18. Satchwell,S.C., Drew,H.R., and Travers,A.A. (1986) *J. Mol. Biol.*, **191**, 659.
19. Hayes,J.J., Tullius,T.D., and Wolffe,A.P. (1990) *Proc. Natl. Acad. Sci. USA*, **87**, 7405.
20. Hayes,J.J., Clark,D.J., and Wolffe,A.P. (1991) *Proc. Natl. Acad. Sci. USA*, **88**, 6829.
21. White,J.H., Gallo,R., and Bauer,W.R. (1989) *J. Mol. Biol.*, **207**, 193.
22. White,J.H., Gallo,R., and Bauer,W.R. (1989) *Nucleic Acids Res.*, **17**, 5827.
23. White,J.H. and Bauer,W.R. (1989) *Cell*, **56**, 9.
24. Hayes,J.J., Bashkin,J., Tullius,T.D., and Wolffe,A.P. (1991) *Biochemistry*, **30**, 8434.
25. Wasserman,S.A., White,J.H., and Cozzarelli,N.R. (1988) *Nature*, **334**, 448.

4

Knots and catenanes

1. Introduction

In Chapter 2, rubber tubing and ribbon were used to model the behaviour of DNA. It is possible with both these materials to make knots and catenanes (a catenane is simply the interlinking of two rings, such as in the links of a chain). It may seem surprising that DNA can also form these structures. This chapter shows that knots and catenanes do indeed occur in DNA and their properties and biological relevance are discussed.

2. Knots

2.1. Occurrence of knots

Knots were first observed in single-stranded DNA by Wang and co-workers in 1976 (1). During studies of the single-stranded circular DNA of bacteriophage fd, it was noted that treatment of the DNA with the enzyme topoisomerase I from *Escherichia coli* (then called ω protein, see Chapter 5), yielded a new species which sedimented faster than the untreated DNA under conditions of high pH, when the double helix was denatured (see Chapter 1, Section 2.1). Examination of this new species by electron microscopy showed that it consisted of DNA rings containing at least three cross-over points. By contrast, the untreated single-stranded circular bacteriophage fd DNA showed very few cross-over points. It was concluded that the fast-sedimenting species were in fact knotted single-strand circles generated by the action of topoisomerase I. It was further shown that the knotted species could be converted back to the unknotted form by the same enzyme. Under high salt conditions *E. coli* topoisomerase I can convert circular bacteriophage fd DNA to a knotted form, whereas under low salt conditions the enzyme catalyses the reverse reaction.

Knots in naturally-occurring double-stranded DNA were first observed in bacteriophage P2 DNA in 1981 by Wang and co-workers (2). Mature bacteriophage P2 DNA consists of double-stranded molecules 33 kb long with 19-base

63

single-strand extensions at each end. In solution these single-strand extensions can base pair to form circular and concatameric (joined end-to-end) species. The DNA found in tailless P2 capsids (heads) is much less viscous than that found in the mature bacteriophage particle. This reduced viscosity has been found to be due to the presence of knotted DNA. Band sedimentation experiments showed that head DNA contains a large fraction of material which sediments more rapidly than the DNA isolated from the mature bacteriophage. Analysis of the head DNA by electron microscopy showed that many of the molecules are highly condensed, consistent with them being in a highly knotted form. The structure of the knotted species is dependent upon hydrogen bonding between the 19-base extensions; disruption of the hydrogen bonding by heating converted the DNA to the linear form. Thus the DNA molecules are knotted, double-stranded DNA rings containing one nick (broken phosphodiester bond) in each strand. These nicks can be sealed by the action of DNA ligase. Sealing of one nick leads to the production of a singly-nicked knotted DNA circle which, under alkaline conditions, dissociates to a single-stranded linear species and a highly knotted single-stranded ring. Sealing of both nicks generates knotted double-stranded circles which cannot be resolved into single-stranded components by alkali treatment.

Naturally-occurring knots in double-stranded DNA have also been found in the tailless capsids of bacteriophage P4 DNA (3). The origin of the knots in P2 and P4 head DNA may be due to random circularization of the DNA in the confined space of the bacteriophage head, where the two ends of the DNA 'find' each other after a random walk through the tangle of DNA in the head. (Imagine the knots which might occur when trying to join the ends of a long piece of string crammed into a restricted space.)

Aside from the discovery of naturally-occurring knotted double-stranded DNA in bacteriophage P2, such knots can also be generated in the test-tube by a variety of enzymes. As already described, the enzyme *E. coli* topoisomerase I can knot single-stranded bacteriophage fd DNA. If double-stranded DNA circles bearing a single-strand nick are used as substrates for this enzyme, knotted DNA species can also be generated (4). These species can be distinguished from unknotted forms by agarose gel electrophoresis and electron microscopy (e.g. *Figure 4.1*). The knotting of intact double-stranded DNA circles can be achieved by the action of type II topoisomerases such as the enzyme from bacteriophage T4 (5). (For a fuller description of the reactions of topoisomerases see Chapter 5.) Knotted double-stranded DNA can also be generated by enzymes involved in recombination reactions (see Chapter 6, Section 6.1).

2.2. Description of knots

The distinction between the general term knot (as applied for example to a piece of string) and a topological knot, is that the latter must be a closed curve. Any knot tied in a piece of string becomes a topological knot when the two ends are joined together. The simplest type of knot that can be made in closed-circular DNA, or any other closed curve, is a trefoil, so-called because the structure when laid flat can be seen to have three lobes (*Figure 4.2a*). The easiest way to

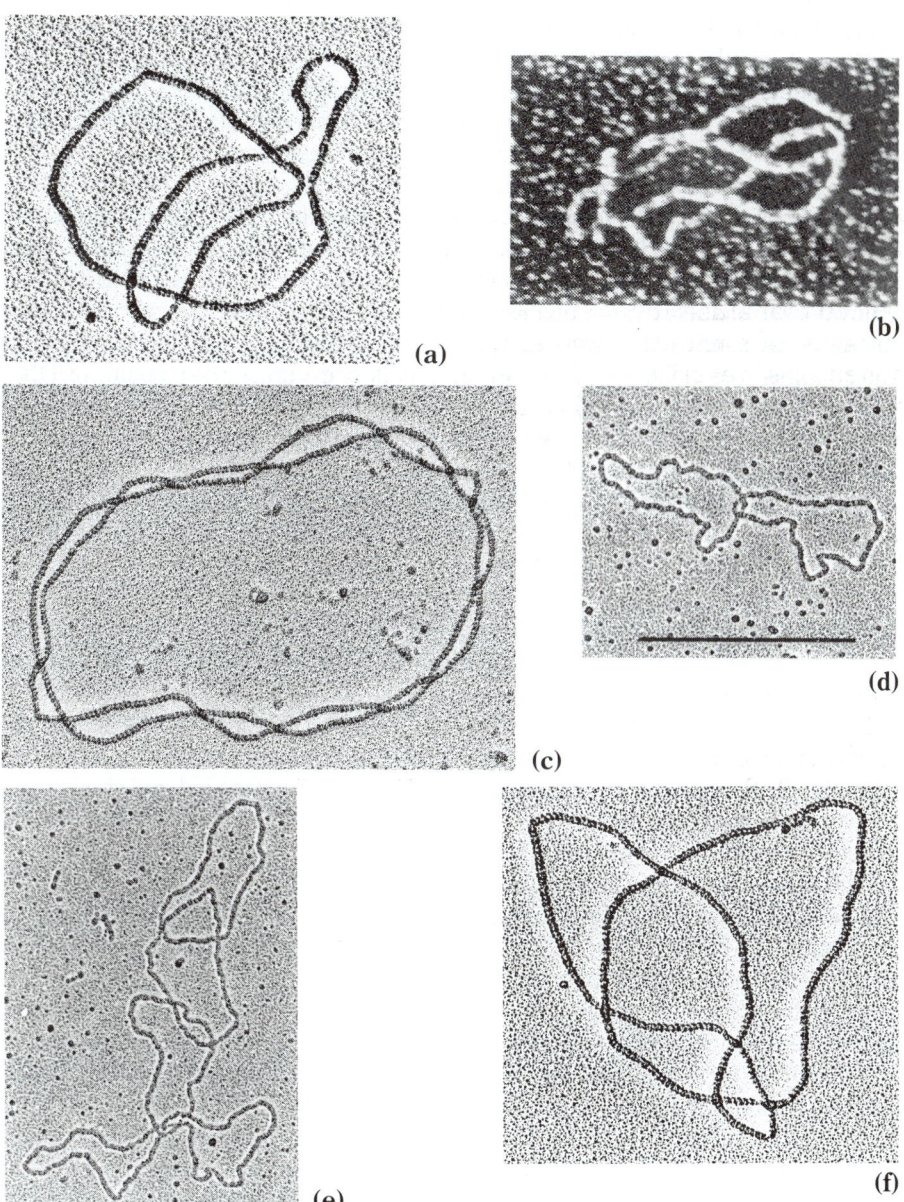

Figure 4.1. Electron micrographs of DNA knots and catenanes. Knotted and catenated DNA were coated with RecA protein prior to visualization by electron microscopy. (a) Trefoil (3-noded knot) (reproduced from ref. 6). (b) Five-noded knot (reproduced from ref. 26). (c) 13-noded torus knot (reproduced from ref. 27). (d) Singly-linked catenane (reproduced from ref. 23). (e) Catenane consisting of five circles (reproduced from ref. 23). (f) Figure-eight (five-noded) catenane (reproduced from ref. 6). (a), (f) reprinted by permission; Copyright © 1983 Macmillan Magazines Limited; (c), (d), (e) Copyright 1980, 1985 Cell Press.

understand this is to take a length of string and pass one end around the other, forming a simple overhand knot, and then secure the ends, either by holding them firmly in one hand or by tying a tight reef knot at the ends. To reveal the trefoil structure it is necessary to lay the string flat on a surface and identify the three lobes, which can be pulled out to resemble the structures shown in *Figure 4.2a*. There are in fact two distinguishable trefoil knots. If the original overhand knot was made in a right-handed (clockwise) sense then the resulting trefoil will be equivalent to that illustrated on the right in *Figure 4.2a*. If the original knot was made in a left-handed (counter-clockwise) sense, the resulting trefoil will be equivalent to that illustrated on the left in *Figure 4.2a*. These two structures are mirror images of each other and are most easily distinguished by the signs of their 'nodes'. As described in Chapter 2, nodes are cross-over points of DNA strands. If a polarity is assigned to the DNA (which in the case of a knot can be completely arbitrary), then the nodes can be given a +1 or −1 designation. When arrows showing the polarity are drawn at the site of the node (as shown in *Figure 4.2a*), the node sign is determined by the direction of movement required to align the arrows on the upper and lower curve with a rotation of the upper arrow of less than 180° (see Chapter 2, Section 3.1). A clockwise rotation defines a −1 node; a counter-clockwise rotation a +1 node. Note that the nodes in Chapter 2 referred to single strands of DNA crossing each other while the nodes described here refer to double strands. However, for knots, these treatments are entirely consistent. If the single strands are drawn at the nodes in *Figure 4.2a*, then four single-strand crossings will occur. Two of these are crossings of one strand over itself and therefore do not contribute to linking (Chapter 2, Section 3.1.1). If the two other nodes are assigned and the sum of the values halved (as described in Chapter 2), the designation of the node will be exactly as given in *Figure 4.2a*.

The above discussion demonstrates the utility of the node convention which can be applied to supercoiling, knotting, and, as described in the next section, to catenation. In DNA the sign of a node can be determined by electron microscopy, a procedure which is greatly facilitated if the DNA is coated with protein, such as the *E. coli* RecA protein (6) (*Figure 4.1*). As *Figure 4.2a* shows, the simplest knot, the trefoil, exists as two stereoisomers, containing either three positive or three negative nodes, when drawn in its most simplified form (*Figure 4.2a*). The sum of the nodes is defined as the intrinsic linkage of the knot, Kn, and is −3 or +3 for the two trefoils (7). Thus the Kn notation adequately distinguishes the two forms of a three-noded knot. The four-noded, or figure-8, knot (*Figure 4.2c*) has two negative and two positive nodes, with $Kn = 0$. In fact the mirror image of this knot is superimposable and there is, rather surprisingly, only one four-noded knot. For knots with five or more nodes there can be different knots having the same value of Kn; thus, Kn does not describe a knot uniquely.

One method for uniquely describing knots (and also catenanes, see Section 3.2) is termed the two-bridge method, which has been described by White and Cozzarelli (8). This involves redrawing the knot in the symmetrical configuration

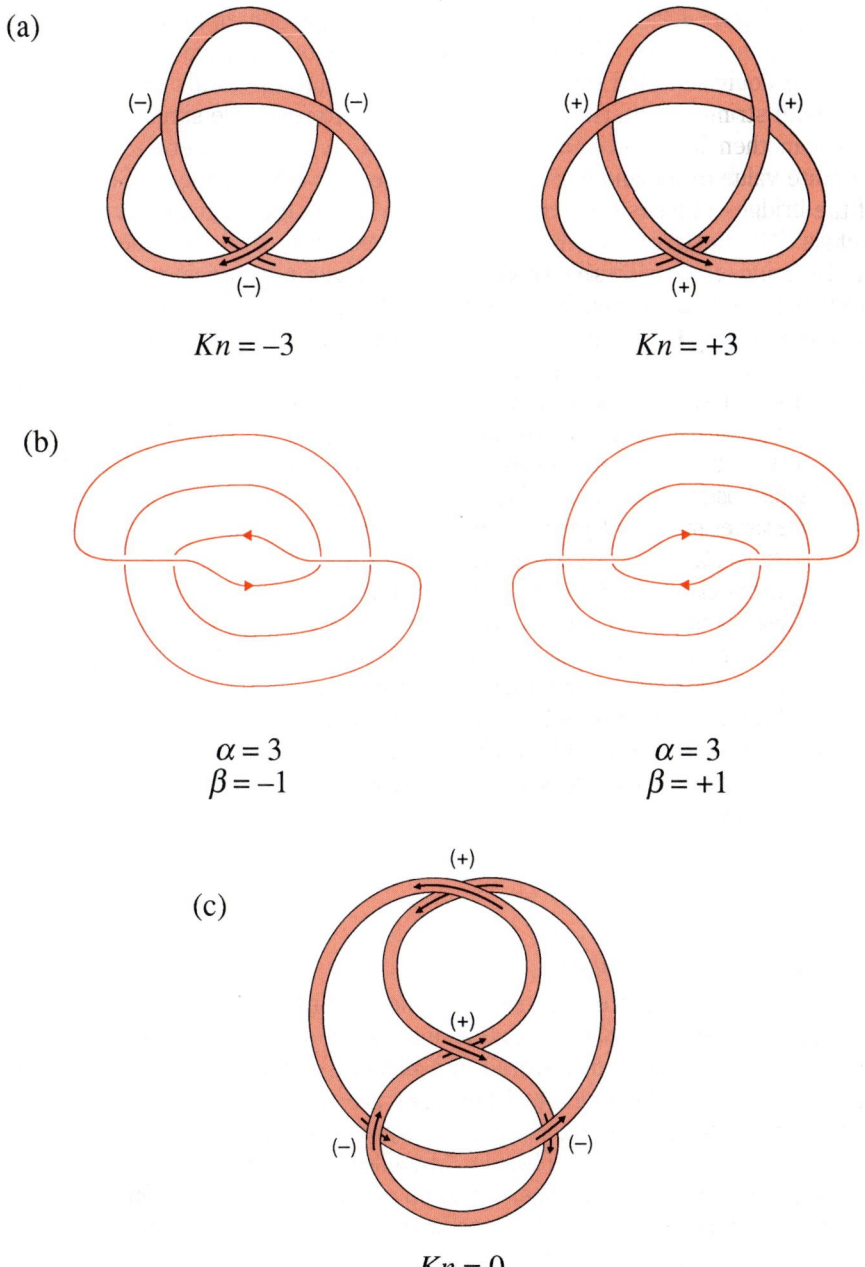

Figure 4.2. Illustration of knots. **(a)** Shows the two isomers of the trefoil (three-noded) knot, which can be redrawn as shown in **(b)** to allow assignment of the parameters α and β, by the two-bridge method (see text). **(c)** Shows the four-noded, or figure-eight knot.

shown in *Figure 4.2b*, where there is a horizontal section, or bridge, on each side of the symmetry axis. Although the forms in *Figure 4.2a* and *b* look dissimilar, they are equivalent and this can be seen by making the forms shown in *Figure 4.2b* with string and rearranging the structures to resemble simple trefoils. The knot can then be described by two integers, α and β, where $\alpha > |\beta|$ (the absolute value of β), and both are odd numbers. α is the number of times either of the bridges crosses the curves, plus one, and is thus equal to three for both trefoils. The designation of β is more complex. First of all the nodes generated by the bridges crossing the curves must be given a sign using the convention previously mentioned. Then the bridge nodes should be numbered as follows. Choose any bridge, move along it in the direction of the orientating arrow; the last node is then given the number 1 and whatever sign has been assigned to that node. Moving backwards along the bridge the other nodes are numbered 2, 3 etc., and given their appropriate signs. Finally, start on the other bridge and move in the direction of the arrow until it crosses under the labelled bridge. The signed number attached to that node is the value of β. Thus for the two trefoils the values of β are $+1$ and -1 (*Figure 4.2b*). Other notations for the description of knots also exist and are discussed elsewhere (8,9).

Knots can, in addition, be classified into two basic types: 'torus' knots and 'twist' knots. Torus knot are characterized by the spiral path of the DNA, which can be thought of as winding around an imaginary torus (cf. Chapter 2, Section 4.2.1). An example of such a knot is shown in *Figure 4.1c*. Torus knots can have only odd numbers of nodes. The other category, twist knots, consist of an interwound region (see Chapter 2, Section 4.2.1) with any number of nodes, and a two-noded interlocking region. An example of a twist knot is the four-noded knot shown in *Figure 4.2c*.

3. Catenanes

3.1. Occurrence of catenanes

Although knots in DNA can be readily generated in the test-tube, they are not particularly common in nature. Catenanes, on the other hand, are rather more common and have been discovered in a number of diverse biological systems; indeed all natural populations of DNA rings are to some extent, interlocked as catenanes. Catenated DNA molecules were first observed in 1967 in the mitochondria of human cells (10,11). Mitochondrial DNA is known to exist as closed-circular duplex molecules. Caesium chloride density gradient centrifugation of mitochondrial DNA in the presence of ethidium bromide produced three bands. The lowest band largely consisted of closed-circular monomer DNA and the top band of topologically unconstrained (nicked or linear) DNA. These two forms have different buoyant densities under these conditions due to the preferential binding of ethidium bromide to topologically unconstrained DNA (see Chapter 2, Section 6.1). The middle band was shown by electron microscopy to contain catenated DNA molecules, in which one molecule was a closed-circular duplex and the other a nicked-circular duplex (10). Such species would be expected to have an intermediate buoyant density.

Catenated DNA has also been found in the mitochondria of trypanosomes. Here the mitochondrial genome consists of a compact network of DNA, known as kinetoplast DNA (kDNA), with a molecular weight of up to 4×10^{10} Da, comprising about 10 000 circles held together by catenation (12). These networks can be visualized by electron microscopy and consist of two types of circular DNA molecules: maxi-circles, 6–12 μm in length (20–38 kb), and mini-circles, 0.2–0.8 μm in length (0.7–2.5 kb), which are linked together into a vast network by catenation. It has been shown that kDNA can be decatenated by type II topoisomerases (see Chapter 5) but not by type I topoisomerases (13), consistent with the kDNA comprising covalently-closed double-stranded circles (see Chapter 5, Section 2).

Catenanes in DNA were first made artificially by circularizing bacteriophage λ DNA in the presence of circular bacteriophage 186 DNA (14). This generated a band on a caesium chloride gradient that was distinct from that derived from either of the bacteriophage DNAs alone. Positive identification of this band as catenated DNA was subsequently made by electron microscopy [e.g. *Figure 4.1*; (15)].

Perhaps the most common manifestation of catenated DNA is as an intermediate during the replication of circular DNA molecules. This has been observed in a variety of systems including bacterial plasmids (16–18) and the animal virus SV40 (19). Catenane formation during replication is one of the potential topological consequences of this process and will be considered further in Chapter 6. DNA catenanes may also be generated by a variety of enzymes in the test-tube. These include the bacteriophage λ Int protein, resolvases, and DNA topoisomerases (20–23). As with unknotting, the DNA topoisomerases can also catalyse the decatenation of catenated DNA rings [see Chapter 5, Section 2 (5,22,23)].

3.2. Description of catenanes

DNA catenanes can be described using the node convention (Section 2.2). However, for two (or more) interlocked DNA rings the polarity is not necessarily arbitrary. For homologous DNA circles, or ones derived from a single ring by recombination, the DNA sequence provides a way to orientate the rings. For circles of different origins the assignment is arbitrary. The simplest type of catenane is a pair of singly interlinked rings as shown in *Figure 4.3a*. There are two forms of such a catenane having either two positive or two negative nodes (*Figure 4.3a*). The easiest way to be convinced of this is to interlink two pieces of rubber tubing joined into circles with connectors. If a direction arrow is drawn on the tubing it can be seen that they can only be interlinked in two ways. As with knots, catenanes can be assigned an intrinsic linkage, Ca, equal to the sum of the intermolecular nodes [*Figure 4.3a*; (7)].

Again, as with knots, Ca does not uniquely define all catenanes, for example, doubly-interlocked catenanes. The two-bridge method can be used to assign α and β values to catenanes (*Figure 4.3b*) as described for knots (8). The torus

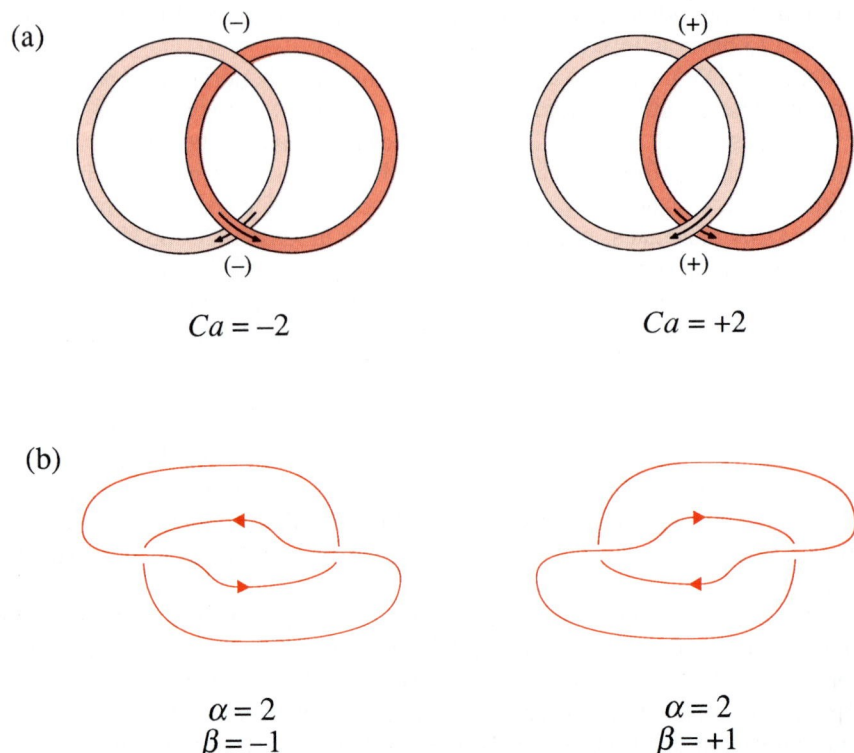

Figure 4.3. Illustration of catenanes. **(a)** Shows the two isomers of the singly-linked catenanes, which can be redrawn as shown in **(b)** to allow assignment of the parameters α and β, by the two-bridge method (see text).

classification of knots also applies to catenanes, where the DNA can be thought of as winding on an imaginary torus. Torus catenanes can only have an even number of nodes. The singly-linked catenanes shown in *Figure 4,3* are examples of torus catenanes. Many other types of catenanes are known (see *Figure 4.1*), including figure-eight catenanes (e.g. Chapter 6, *Figure 6.7*), complex multiply-linked catenanes, and catenanes involving many DNA circles, such as those found in the kinetoplast DNA of trypanosomes (see Section 3.1).

4.   Knots and catenanes as probes of DNA–protein interactions

The use of knots and catenanes in the study of site-specific recombination reactions of bacteriophage λ Int protein and Tn3 resolvase will be discussed in detail in Chapter 6, Section 6, but it is worth emphasizing their usefulness by discussing here some examples from other systems.

Enhancers are DNA sequences that increase the rate of transcription of genes. These elements occur in both eukaryotes and prokaryotes, and are thought to serve as protein binding sites. They are characterized by their ability to function over large distances. Models of their mechanism of action include those where communication between the enhancer and the promoter is mediated by the DNA between them, for example, by the 'tracking', or sliding of a protein along the intervening DNA (see Chapter 6, Section 6.2). One test of such models involved a singly-linked catenane, where one circle contained an enhancer, and the other a promoter (24). It was found that transcription from the promoter was still stimulated despite the elements being on different DNA rings; separation of the two rings by decatenation abolished stimulation. This experiment shows that the enhancer and promoter need not be contiguous on the DNA, but do need to be able to be brought into close proximity, favouring DNA looping (see Chapter 6, Section 4.2) as a mechanism of transcriptional activation.

Bacteriophage Mu encodes a site-specific recombination system whereby a protein (Gin) inverts a region of Mu DNA between two recombination sites (known as *gix*), enabling expression of alternative tail fibre genes, each conferring a particular host range on the bacteriophage. Gin is a member of a family of proteins termed DNA invertases, which share many similarities with the resolvase proteins (see Chapter 6, Section 6.2). Experiments with knotted substrates, and substrates with the *gix* sites present on distinct, but multiply-linked catenated DNA molecules, have shown that sites can efficiently recombine when on separate molecules, thus eliminating tracking mechanisms, as before. In addition, the results suggest the coming together of the *gix* sites, and a further site involved in promoting the reaction, in a complex whose structure utilizes the right-handed wrapping in interwound (plectonemic) supercoiled DNA (see Chapter 2, Section 4.2.1). This plectonemic conformation can be mimicked by knotting or catenation in the substrate molecules (25).

5. Conclusions

Knots and catenanes might appear to be esoteric structures and their descriptions can involve fairly complex mathematical concepts. Nevertheless, they are found in nature, and their origins may well reflect important features of the reactions of DNA-specific enzymes. Indeed, careful analyses of knots and catenanes generated by recombination enzymes have yielded important mechanistic information concerning these enzymes (see Chapter 6, Section 6).

6. Further reading

Frisch, H.L. and Klempner, D. (1970) Topological isomerism and macromolecules. *Adv. Macromolec. Chem.*, **2**, 149.

Wasserman, S.A. and Cozzarelli, N.R. (1985) Determination of the stereostructure of the product of Tn3 resolvase by a general method. *Proc. Natl. Acad. Sci. USA*, **82**, 1079.

Wasserman,S.A. and Cozzarelli,N.R. (1986) Biochemical topology: applications to DNA recombination and replication. *Science*, **232**, 951.

Watson,A. (1991) Twists, tangles and topology. *New Scientist*, **5 October**, 42.

7. References

1. Liu,L.F., Depew,R.E., and Wang,J.C. (1976) *J. Mol. Biol.*, **106**, 439.
2. Liu,L.F., Perkocha,L., Calendar,R., and Wang,J.C. (1981) *Proc. Natl. Acad. Sci. USA*, **78**, 5498.
3. Liu,L.F. and Davis,J.L. (1981) *Nucleic Acids Res.*, **9**, 3979.
4. Brown,P.O. and Cozzarelli,N.R. (1981) *Proc. Natl. Acad. Sci. USA*, **78**, 843.
5. Liu,L.F., Liu,C.-C., and Alberts,B.M. (1980) *Cell*, **19**, 697.
6. Krasnow,M.A., Stasiak,A., Spengler,S.J., Dean,F., Koller,T., and Cozzarelli, N.R. (1983) *Nature*, **304**, 559.
7. Cozzarelli,N.R., Krasnow,M.A., Gerrard,S.P., and White,J.H. (1984) *Cold Spring Harbor Symp. Quant. Biol.*, **49**, 383.
8. White,J.H and Cozzarelli,N.R. (1984) *Proc. Natl. Acad. Sci. USA*, **81**, 3322.
9. White,J.H., Millet,K.C., and Cozzarelli,N.R. (1987) *J. Mol. Biol.*, **197**, 585.
10. Hudson,B. and Vinograd,J. (1967) *Nature*, **216**, 647,
11. Clayton,D.A. and Vinograd,J. (1967) *Nature*, **216**, 652.
12. Englund,P.T., Hajduk,S.L., and Marini,J.C. (1982) *Annu. Rev. Biochem.*, **51**, 695.
13. Marini,J.C., Miller,K.G., and Englund,P.T. (1980) *J. Biol. Chem.*, **255**, 4976.
14. Wang,J.C. and Schwartz,H. (1967) *Biopolymers*, **5**, 953.
15. Martin,K. and Wang,J.C. (1970) *Biopolymers*, **9**, 503.
16. Kupersztoch,Y.M. and Helsinki,D.R. (1973) *Biochem. Biophys. Res. Commun.*, **54**, 1451.
17. Novick,R.P., Smith,K., Sheehy,R.J., and Murphy,E. (1973) *Biochem. Biophys. Res. Commun.*, **54**, 1460.
18. Sakakibara,Y., Suzuki,K., and Tomizawa,J.-I. (1976) *J. Mol. Biol.*, **108**, 569.
19. Jaenisch,R. and Levine,A.J. (1973) *J. Mol. Biol.*, **73**, 199.
20. Mizuuchi, K., Fisher,L.M., O'Dea,M.H., and Gellert,M. (1980) *Proc. Natl. Acad. Sci. USA*, **77**, 1847.
21. Reed,R.R. (1981) *Cell*, **25**, 713.
22. Tse,Y.-C. and Wang,J.C. (1980) *Cell*, **22**, 269.
23. Kreuzer,K.N. and Cozzarelli,N.R. (1980) *Cell*, **20**, 245.
24. Wedel,A., Weiss,D.S., Popham,D., Dröge,P., and Kustu,S. (1990) *Science*, **248**, 486.
25. Kanaar,R., van de Putte,P., and Cozzarelli,N.R. (1989) *Cell*, **58**, 147.
26. Shishido,K., Komiyama,N., and Ikawa,S. (1987) *J. Mol. Biol.*, **195**, 215.
27. Spengler,S.J., Stasiak,A., and Cozzarelli,N.R. (1985) *Cell*, **42**, 325.

5

DNA topoisomerases

1. Introduction

It is true to say that virtually every reaction that occurs in biological systems is catalysed by an enzyme. The interconversions of different topological forms of DNA are no exceptions. The enzymes which can catalyse these processes are known as DNA topoisomerases and they constitute a widespread and interesting group of proteins. Every cell type examined so far has been found to contain DNA topoisomerases, and, where a genetic test has been possible, it can be shown that at least one is essential for cell growth. Examples of the types of organisms found to contain topoisomerases include the bacteria *Escherichia coli* and *Staphylococcus aureus*, yeasts, *Drosophila*, cauliflower, and man. In addition, several viruses are known to encode a topoisomerase, for example, bacteriophage T4 and the animal virus vaccinia. Examples of DNA topoisomerases are given in *Table 5.1*. Perhaps the most important aspect of topoisomerases for those interested in DNA topology is the mechanism of topoisomerase action: how do these enzymes achieve the seemingly exotic reactions of DNA supercoiling, knotting, and decatenation. The current understanding of how these enzymes work is summarized later in this chapter.

2. Reactions of topoisomerases

The reaction that is common to all topoisomerases isolated so far is the ability to relax negatively supercoiled DNA, that is to convert it into a less supercoiled form. A number of other reactions are also known and these are illustrated in *Figures 5.1* and *5.2*. During studies of these reactions it became clear that topoisomerases could be divided into two types. Type I enzymes are able to carry out reactions involving the breaking of only one strand of the DNA (*Figure 5.1*), while type II enzymes can carry out reactions involving the breaking of both strands of the DNA (*Figure 5.2*). Perhaps the best illustrations of this distinction are the catenation and decatenation of double-stranded DNA circles. These

Table 5.1 DNA topoisomerases

Enzyme	Type	Source	Subunit size (kDa)	Remarks
Bacterial topoisomerase I (ω protein)	I	Bacteria (e.g. *E. coli*)	97	Will not relax positive supercoils
Int protein	I	Bacteriophage λ	40	Proteins normally involved in
Resolvase	I	Transposons (Tn3 family)	21	recombination reactions but which also display topoisomerase activity
Eukaryotic topoisomerase I	I	Eukaryotes (e.g. human)	91	Will relax both positive and negative supercoils
Vaccinia virus topoisomerase I	I	Vaccinia virus	37	ATP stimulates topoisomerase activity
Topoisomerase III[a]	I	Bacteria (*E. coli*)	74	Potent decatenating activity
Reverse gyrase	I	Thermophilic bacteria (e.g. *Sulfolobus*)	128	Will introduce positive supercoils into DNA
DNA gyrase	II	Bacteria (e.g. *E. coli*)	97 and 90	Will introduce negative supercoils in DNA
T4 topoisomerase	II	Bacteriophage T4	58, 51, and 18	ATP-dependent, but can only relax, not supercoil DNA
Eukaryotic topoisomerase II	II	Eukaryotes (e.g. human)	174	
Toposiomerase IV[a]	II	Bacteria (*E. coli*)	67 and 81	

[a] Note that topoisomerases III and IV do not represent new 'types' of topoisomerase mechanism (cf. type I and type II).

reactions can be achieved by type II topoisomerases but not by type I enzymes, unless one of the duplexes already contains a break in one strand (1). A consequence of this difference in mechanism between type I and type II topoisomerases is that the reactions of the former occur in steps with linking number changes of one, while those of the latter occur with linking number changes of two.

Not all topoisomerases can carry out the full range of reactions shown in *Figures 5.1* and *5.2*. For example, the type I topoisomerase from *E. coli* can only relax negatively supercoiled DNA while the type I enzyme from calf thymus will relax both negatively and positively supercoiled DNA (2,3). This turns out to be a consequence of the inability of *E. coli* topoisomerase I to bind to positively supercoiled DNA. A further example concerns the ability of topoisomerases to introduce supercoils into DNA. Only one type II enzyme, called DNA gyrase, is able to do this. DNA gyrase is found in bacteria and is able to introduce negative supercoils into DNA using the free energy from ATP hydrolysis (4). In

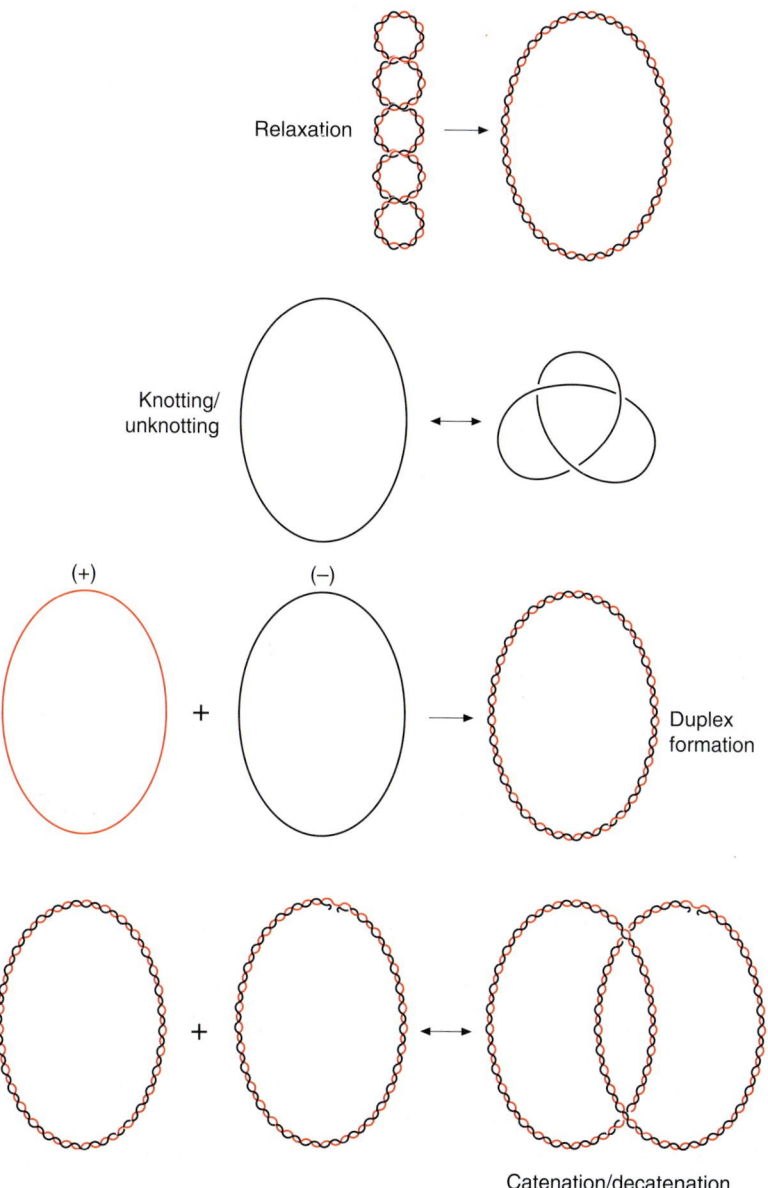

Catenation/decatenation

Figure 5.1. The reactions of type I topoisomerases (redrawn from ref. 8).

thermophilic bacteria such as *Sulfolobus* a second gyrase exists, called reverse gyrase, which can introduce positive supercoils into DNA (5). Type II enzymes from eukaryotes and bacteriophage are unable to catalyse the introduction of DNA supercoiling.

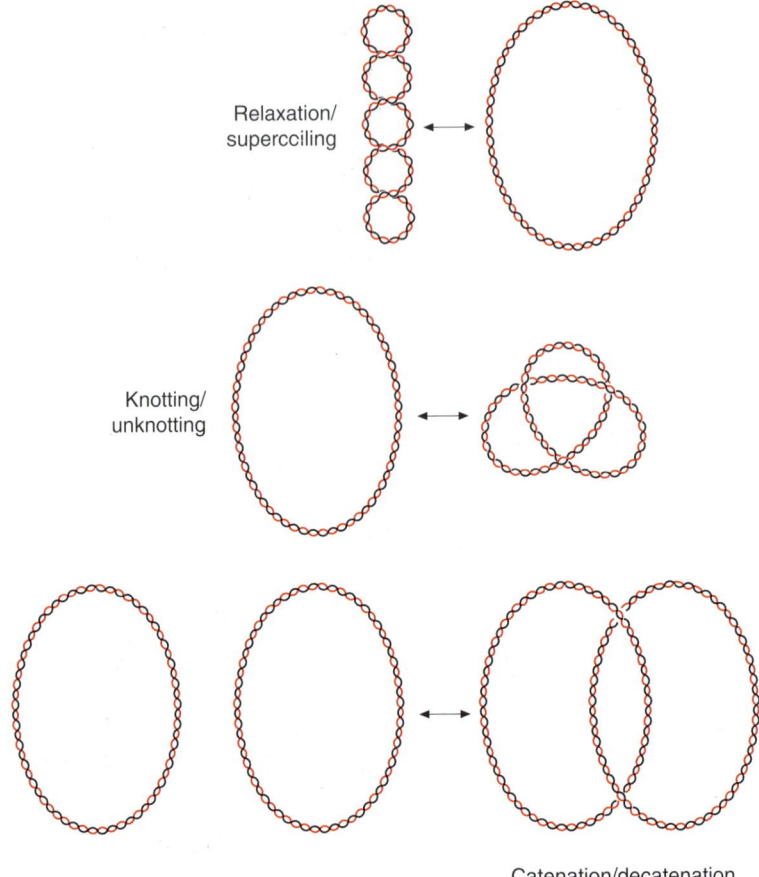

Relaxation/
supercciling

Knotting/
unknotting

Catenation/decatenation

Figure 5.2. The reactions of type II topoisomerases (redrawn from ref. 8).

In addition to topoisomerase I and topoisomerase II (DNA gyrase), *E. coli* has also been found to encode two further topoisomerases (*Table 5.1*). Topoisomerase III is another type I enzyme which has been found to be more active as a decatenating enzyme than as a DNA relaxing enzyme (6). Topoisomerase IV is a type II enzyme with a high degree of homology to DNA gyrase, which can relax, but not supercoil, DNA (7). It is likely that multiple topoisomerases will also be found in other species.

There are now a large number of enzymes which have been shown to perform reactions of the type shown in *Figures 5.1* and *5.2*. Some of these were not originally identified as topoisomerases but as enzymes involved in recombination reactions. Examples include resolvase proteins involved in the process of transposition and the Int protein involved in bacteriophage λ integration (see Chapter 6, Section 4). These proteins carry out reactions involving the breaking of DNA and the transfer of the broken end to another DNA molecule. As will be seen, this is mechanistically very similar to a topoisomerase reaction except that topoisomerases rejoin the broken end to its original site.

3. Mechanisms of topoisomerases

Considering the DNA relaxation reaction in the light of what is known about linking number changes in DNA from Chapter 2, it would seem logical that topoisomerases should work by a swivel mechanism. This would involve breaking one (or both) strands of the DNA, allowing the free end (or ends) to rotate about the helix axis, and resealing of the break. Such a process would alter the linking number of the DNA as required by the relaxation reaction. However, consideration of the knotting/unknotting and catenation/decatenation reactions suggests that a swivel mechanism can not account for the full range of topo-isomerase reactions. In fact a different type of mechanism, called strand passage, can account for the ability of topoisomerases to catalyse all these reactions (8). In its simplest form, strand passage involves the cleavage of one or both strands of the DNA by the enzyme and the passing of a single- or double-stranded segment of DNA through the break which is then resealed. It has been found that topoisomerases stabilize the DNA break by forming a covalent bond between the enzyme (usually via a tyrosine hydroxyl group) and the phosphate at the break site. The strand passage event can involve segments of DNA from the same DNA molecule, in the case of relaxation/supercoiling and knotting/unknotting, or from separate DNA molecules, in the case of catenation/decatenation. The details of the strand passage mechanism differ from one enzyme to another. The reactions of type I enzymes proceed via single-strand breaks in DNA and involve covalent attachment to either the 5'-phosphate (prokaryotic enzymes) or the 3'-phosphate (eukaryotic enzymes) at the break site. The reactions of type II enzymes proceed via double-strand breaks and involve covalent attachment at the 5'-phosphate. Examples of the proposed mechanisms for certain topoisomer-ases are shown in *Figures 5.3* and *5.4*.

Perhaps the most remarkable of the topoisomerase mechanisms is that of the bacterial type II topoisomerase DNA gyrase (*Figure 5.4*), which involves the coupling of ATP hydrolysis to the strand passage process (9). DNA supercoiling is an energetically unfavourable process and gyrase is somehow able to transform the chemical energy derived from the hydrolysis of a phosphodiester bond in ATP into the torsional stress of supercoiling. Although it must be assumed that this process involves conformational changes in the protein, nothing is known of the detailed mechanism of this energy coupling process. It is interesting to note that other type II topoisomerases hydrolyse ATP but are able only to relax, not supercoil DNA; *Table 5.1*. Gyrase can also relax both positively and negatively supercoiled DNA; the former reaction requires ATP. Another noteworthy feature of the gyrase reaction is the wrapping of DNA around the protein. This is reminiscent of other systems in which DNA is wrapped around proteins, such as the nucleosome and RNA polymerase. One proposition is that the

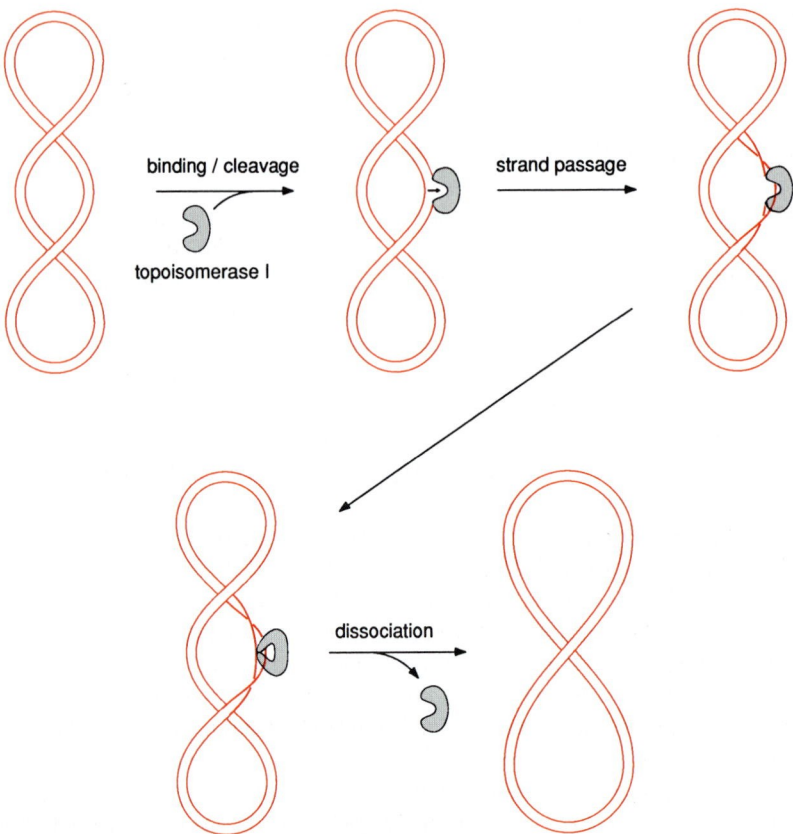

Figure 5.3. A proposed mechanism for the bacterial type I topoisomerase, *E. coli* topoisomerase I. The enzyme (shown in orange) binds to DNA and cleaves one strand. This single-stranded break is stabilized by the formation of a DNA–protein covalent bond. The intact strand is passed through the break which may then be resealed. This results in the relaxation of one negative supercoil ($\Delta Lk = +1$).

handedness of this wrap (right-handed) in gyrase determines the directionality of the supercoiling reaction, that is, towards negative supercoiling. The 'strand passage' mechanism of type II topoisomerases involves the translocation of DNA both through the double-stranded break and through the protein complex. Crystallographic evidence suggests that for DNA gyrase this is facilitated by the existence of a 'hole' between two of the subunits (10).

4. Topoisomerases as drug targets

In the opening section of this chapter it was noted that topoisomerases have been found to be essential for cell growth. As a consequence these enzymes are

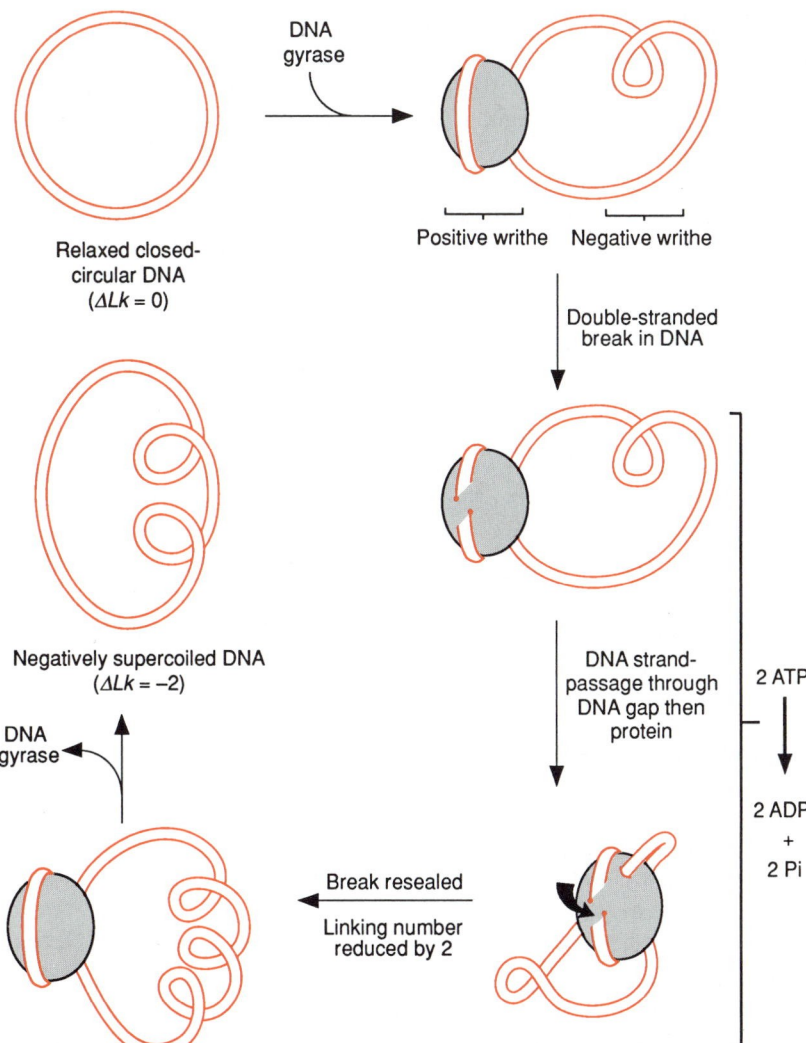

Figure 5.4. A proposed mechanism for the bacterial type II topoisomerase, *E. coli* DNA gyrase. The enzyme (shown in orange) binds to DNA and stabilizes a positive writhe, resulting in the formation of a negative writhe elsewhere in the DNA. The wrapped DNA is cleaved in both strands by the protein and the double-strand break is stabilized by DNA–protein covalent bonds. Another segment of the DNA is then passed through this break (and then through the enzyme complex); the break may then be resealed. This results in the introduction of two negative supercoils ($\Delta Lk = -2$). Catalytic supercoiling requires the hydrolysis of ATP and it is believed that two ATPs are hydrolysed per catalytic cycle.

potential targets for cytotoxic drugs (11,12). For example, DNA gyrase is the target for two groups of anti-bacterial compounds, the quinolones and the coumarins. In the context of topoisomerase mechanisms it is interesting to consider how these drugs might work. The quinolones (e.g. nalidixic acid and ciprofloxacin) are thought to act by interrupting the step in the supercoiling reaction in which the double-strand break is resealed (*Figure 5.4*)). Indeed it can be shown that, under certain conditions, quinolones can lead to trapping of the gyrase–DNA intermediate in which the A subunits of the enzyme are covalently bound to the 5′-phosphates of DNA at the break site. The coumarin drugs (e.g. novobiocin and coumermycin A_1) act in an entirely different way by inhibiting the hydrolysis of ATP by gyrase and thus preventing DNA supercoiling.

The eukaryotic topoisomerases have also been shown to be the targets of drugs. The anti-tumour drug camptothecin has been shown to act on eukaryotic topoisomerase I. The drug appears to work by inhibiting the DNA resealing step of the topoisomerase reaction possibly in a similar manner to the action of quinolones on DNA gyrase. Several other anti-tumour drugs have been shown to be inhibitors of eukaryotic topoisomerase II. These include acridines (e.g. amsacrine), ellipticines (e.g. 2-methyl-9-hydroxy-ellipticinium acetate) and epipodophyllotoxins (e.g. teniposide). Again these drugs are thought to act by inhibiting the DNA resealing step and in some cases this is thought to be mediated by intercalation of the drug into the DNA at the site of enzyme binding.

5. Biological role of topoisomerases

As will be discussed in detail in Chapter 6, changes in DNA topology occur in many cellular processes and have profound biological consequences. Therefore it is not surprising to find that DNA topoisomerases are directly or indirectly involved in these processes.

DNA replication is at the heart of a cell's viability. Topoisomerases are thought to be involved in various stages of this process. DNA gyrase has been shown to be important for the initiation of replication in prokaryotes. Whether this requirement reflects a specific role or merely the need for negative supercoiling prior to initiation is not clear. During the elongation steps of replication the parental DNA is continually being unwound and topoisomerases are required to prevent the accumulation of positive supercoils. Both type I and type II topoisomerases have been shown to be capable of relieving the torsional stress generated during elongation. At the termination of DNA replication the progeny DNA molecules are frequently found to be catenated and it has been shown that topoisomerases can resolve the intertwined molecules.

Unlike replication, transcription could theoretically proceed without any topological problems. However, it has been shown in bacteria that transcription can lead to the supercoiling of the DNA, when, for example, the DNA is anchored to fixed points or when two genes on a circular plasmid are being transcribed in opposite directions (see Chapter 6, Section 3). Transcription can lead to positive supercoiling ahead of the transcription complex and negative supercoiling be-hind it. Experiments have suggested that DNA gyrase can relax the positive supercoils and topoisomerase I can relax the negative supercoils (13). For example, highly positively supercoiled plasmids have been isolated from *E. coli* treated with coumarin drugs (which inhibit DNA gyrase), and plasmids isolated from *E. coli* strains carrying mutations in *topA* (the gene encoding topoisomerase I) can exhibit high levels of negative supercoiling (see Chapter 6, Section 3).

An important function of topoisomerases within the cell is the maintenance of DNA supercoiling. It is known that in bacteria the level of intracellular supercoil-ing influences the rate of transcription of many genes. Given that RNA polymer-ase unwinds DNA upon binding to promoters it is to be expected that negative supercoiling should stimulate transcription. In fact negative supercoiling can both increase and decrease the expression of genes. The topoisomerase genes them-selves are affected by supercoiling. Lowering negative supercoiling raises the expression of the gyrase genes and reduces the expression of *topA*. It is thought that this represents a homeostatic mechanism of the control of DNA supercoiling within the bacterial cell (14).

6. Conclusions

The existence of an essential class of enzymes whose function is to interconvert different topological isomers of DNA, demonstrates the vital importance of DNA topology in cells. Apart from their relevance to DNA topology, topoisomerases are of great interest from the standpoint of their mechanistic enzymology and because of their potential as drug targets.

7. Further reading

Austin,C.A. and Fisher,L.M. (1990) DNA topoisomerases: enzymes that change the shape of DNA. *Sci. Prog. Oxford*, **74**, 147.

Cozzarelli,N.R. and Wang,J.C. (1990) DNA topology and its biological effects. Cold Spring Harbor Laboratory Press, Cold Spring Harbor.

Drlica,K. (1990) Bacterial topoisomerases and the control of DNA supercoiling. *Trends Genet.*, **6**, 433.

Wang,J.C. (1982) DNA topoisomerases. *Sci. Am.*, **247 (July)**, 84.

8. References

1. Liu,L.F., Liu,C.-C., and Alberts,B.M. (1980) *Cell*, **19**, 697.
2. Champoux,J.J. and Dulbecco,R. (1972) *Proc. Natl. Acad. Sci. USA*, **69**, 143.
3. Wang,J.C. (1971) *J. Mol. Biol.*, **55**, 523.
4. Gellert,M., Mizuuchi,K., O'Dea,M.H., and Nash,H.A. (1976) *Proc. Natl. Acad. Sci. USA*, **73**, 3872.
5. Kikuchi,A. and Asai,K. (1984) *Nature*, **309**, 677.

6. DiGate,R.J. and Marians,K.J. (1988) *J. Biol. Chem.*, **263**, 13366.

7. Kato,J., Nishimura,Y., Imamura,R., Niki,H., Hiraga,S., and Suzuki,H. (1990) *Cell*, **63**, 393.

8. Maxwell,A. and Gellert,M. (1986) *Adv. Prot. Chem.*, **38**, 69.

9. Reece,R.J. and Maxwell,A. (1991) *CRC Crit. Rev. Biochem. Mol. Biol.*, **26**, 335.

10. Wigley,D.B., Davies,G.J., Dodson,E.J., Maxwell,A., and Dodson,G. (1991) *Nature*, **351**, 624.

11. Drlica,K. and Franco,R.J. (1988) *Biochemistry*, **27**, 2253.

12. Liu,L.F. (1989) *Annu. Rev. Biochem.*, **58**, 351.

13. Liu,L.F. and Wang,J.C. (1987) *Proc. Natl. Acad. Sci. USA*, **84**, 7024.

14. Menzel,R. and Gellert,M. (1983) *Cell*, **34**, 105.

6

Biological consequences of DNA topology

1. Introduction: the ubiquity of DNA supercoiling

DNA supercoiling is an attribute of almost all DNA *in vivo*. Plasmids, bacterial chromosomes, mitochondrial, and chloroplast DNA, and many viral genomes occur as closed-circular DNA, and as has been described in Chapter 2, linking number is an inherent property of such molecules. Eukaryotic chromosomes and other DNAs such as some yeast plasmids, although consisting of linear DNA, appear to be anchored to a nuclear matrix (scaffolding) at a number of sites, and the domains between such attachment sites behave in a topological sense as closed-circular loops, since their ends are in a fixed orientation relative to one another (1). Such closed-circular molecules or closed domains are almost invariably negatively supercoiled, equivalent to an unwinding of the right-handed DNA helix, the only exceptions being in extremely thermophilic archæbacteria and eubacteria, where plasmids have been isolated with positive supercoiling. As has been seen in Chapter 5, these bacteria also contain a 'reverse gyrase', which is able to introduce positive supercoiling into a relaxed DNA substrate. It seems most likely that the presence of positively supercoiled DNA in such species is an adaptation to the extreme conditions in which they live; positively supercoiled DNA is likely to be more resistant to the unwinding and denaturation of DNA which would be expected at very high temperatures.

1.1. General biological consequences of negative supercoiling

The first and most often quoted consequence of DNA supercoiling is that it allows its compaction into a very small volume. Chromosomal DNA molecules are surprisingly long in relation to the size of bacteria and eukaryotic nuclei; for example, the *Escherichia coli* chromosome is around 1.5 mm long, whilst the cell has a diameter of less than 1 μm. The problem is more acute in eukaryotes where chromosomes of the order of metres in length must be contained within a nucleus also less than 1 μm across. Boles *et al.* (2), however, point out that plectonemic supercoils, the conformation adopted by supercoiled free DNA, are very inefficiently compacted (~2.5-fold) Toroidal winding (see Chapter 2, Section 4.2) is much more efficient, and in eukaryotes at least, it is the binding of DNA in nucleosomes and high-order structures (see Section 2.1) which provides the major compacting effect.

More important, perhaps, is that DNA supercoiling has a direct influence on many DNA-associated processes *in vivo*, which, for the most part, involve the interaction of specific proteins with DNA. Many DNA–protein interactions can be notionally divided into a number of discrete steps which individually may be affected by DNA supercoiling. These include the binding of proteins to the DNA, the bringing together of two or more sites on the DNA ('synapsis') and, in the case of recombinases and topoisomerases, strand-transfer processes.

The binding of proteins to DNA is often supercoiling dependent. Negatively supercoiled DNA (that most frequently found in nature) is in a high energy conformation compared with unconstrained DNA (see Chapter 2, Section 6). This excess energy may be relieved by protein binding. The unfavourable deformations associated with negative supercoiling are the untwisting of the DNA helix and the negative writhing of the helix axis. It follows that any binding process which requires or makes use of these distortions will be favoured by negative supercoiling. The most obvious examples are the binding of proteins which require the unwinding of the DNA helix (such as those involved in DNA replication and transcription). In addition, a number of DNA–protein complexes involve the wrapping of DNA around the protein [e.g. nucleosomes (Section 2.1) and bacteriophage λ Int protein (Section 6.1)], and their binding is promoted by negative supercoiling, due to the stabilization of writhing.

The activities of a number of DNA-specific proteins involve synapsis [e.g. certain transcriptional activators and repressors (Section 4.2), site-specific re-combination proteins (Section 6)]. Plectonemic, or interwound, supercoiling (Chapter 2, Section 4.2.1) provides a way of bringing two distant sites on DNA together (3). If a conveyor belt motion of the DNA around a fixed superhelical axis occurs in plectonemically supercoiled DNA (a process that has been termed 'slithering'), then one can imagine the motion bringing remote sites into apposition, thus facilitating synapsis. Such a mechanism has been suggested for the synapsis of sites involved in recombination by resolvase (4).

Strand-transfer reactions are features of recombinases and topoisomerases (see Chapter 5). Often, these reactions lead to changes in linking number of the DNA substrate. Clearly, if the reaction occurs on negatively supercoiled DNA, then processes leading to a positive linking difference (ΔLk), such as DNA relaxation by topoisomerases, will be favoured. The processes of binding, synapsis, and strand-transfer are general phenomena which are found in many systems of DNA–protein interaction. Section 6 discusses these processes in detail in re-lation to site-specific recombination.

The excess free energy associated with supercoiling, referred to previously, also influences the formation of Z-DNA, cruciforms, and H-DNA structures (see Chapter 1, Sections 2.4, 3.1, and 3.2). The formation of Z-DNA has the effect of dramatically reducing the twist of the DNA, since it converts a section of DNA from right-handed (+ve) to left-handed (−ve) twist. So, for example, if 20–25 bp of B-DNA are converted to the Z form, the twist of the DNA is reduced by about 4. If the linking difference of a 4000 bp DNA plasmid, is −25 (a typical value for DNA isolated from cells), almost a sixth of that ΔLk is concentrated into 0.005 of the plasmid by the formation of the Z-DNA region, so the effective specific linking difference (σ) of the remainder is much lower. This saving in energy compensates for the positive free energy associated with the formation of the more unstable structure and its junctions with normal B-DNA. In the same way, the formation of a cruciform from a given region of DNA is equivalent to the complete unwinding of that region, so the unfavourable effect of the unpairing of base pairs in the loop of the cruciform is offset by the reduction in σ in the rest of the plasmid; much the same argument applies to H-DNA. In general, a certain threshold specific linking difference is required to facilitate isomerization to an alternative structure (5).

Additionally, the helical structure of DNA and its closed-circular nature cause a number of complications in processes such as replication and transcription which require specific mechanisms for their resolution. The consequences of negative supercoiling and other DNA topological forms for DNA-associated processes will be considered in more detail in the following sections.

2. Genome organization

2.1. Eukaryotes

As already mentioned, the genomic DNA in cells must be highly compacted in order to be contained in the space required. In eukaryotes, the first stage of this compaction is the winding of the DNA in the nucleosome, and the resulting histone-associated DNA is referred to as chromatin.

The nature of the wrapping of DNA in the nucleosome has been considered in detail in Chapter 3, Section 5. Briefly, about 146 bp of DNA are wrapped in approximately two left-handed superhelical turns around a protein octamer consisting of histones which are highly basic, positively charged proteins which help to neutralize the negatively charged phosphate backbone of the DNA. This basic unit is repeated many times in eukaryotic chromatin; indeed it seems likely that almost all the DNA is constrained in this way. It follows from this that the negative supercoiling of the DNA is largely partitioned into writhing of the helix axis stabilized by interaction with the histones, and hence is not directly available for promoting the binding of ligands which unwind the DNA helix (see Section 1.1). It is possible that dissociation of DNA from the histone octamer has a part to play in the initiation of processes such as replication and transcription (6–9). The unconstrained supercoiling resulting from this dissociation could be harnessed to promote the untwisting of the DNA helix.

There are several increasing levels of compaction which have been identified in chromatin, in addition to the nucleosome itself. The first is the nucleosome filament, an extended array of nucleosomes along a DNA length, which also

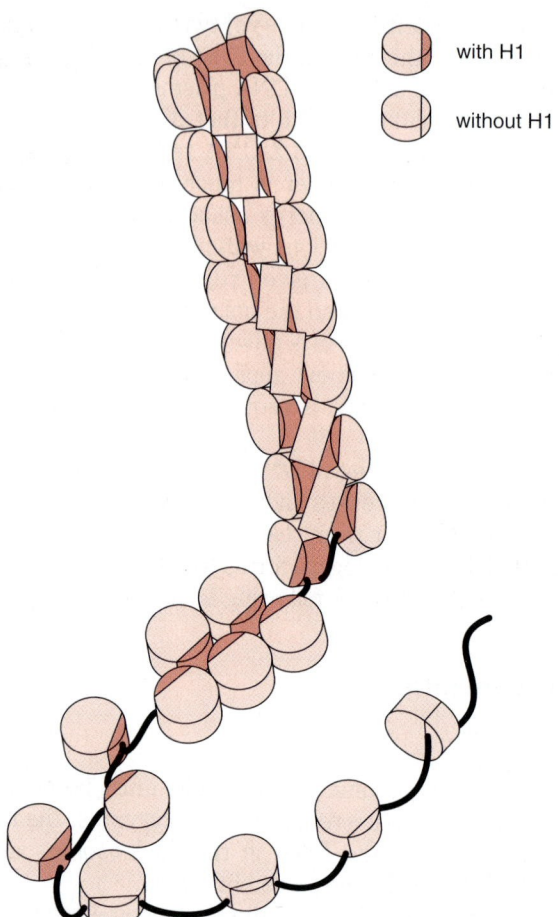

with H1

without H1

Figure 6.1. A model for chromatin condensation. An idealized drawing of higher-order structures formed by chromatin containing histone H1. The open zigzag (bottom left) forms helices with increasing numbers of nucleosomes per turn. The final solenoid probably has about six nucleosomes per turn (redrawn from ref. 12 by permission of the Company of Biologists Ltd.).

contains histone H1, a protein associated with the entry and exit points of DNA from the histone octamer (this filament has a zigzag appearance in electron micrographs) (10). In the 30 nm filament or solenoid, nucleosomes are wrapped into a compact left-handed helix of diameter 30 nm [*Figure 6.1*; reviewed in (11)]. These forms have been fairly well characterized, and it is thought that higher orders of compaction, for example a helical winding of the 30 nm filament, also occur. The level of compaction exhibited by chromatin is dependent on the stage of the cell cycle, for example chromosomes are highly compacted at mitosis, but are more extended during interphase, consisting mostly of the

30 nm fibre (12). In addition, chromatin which is being actively transcribed has a less condensed structure than inactive chromatin, characterized by the loss of histone H1 and most higher-order structure, although many nucleosomes are still present (8,9).

2.2. Prokaryotes

The organization of the chromosome in prokaryotic cells is less clearly defined. The best studied example is the $E.$ $coli$ genome, which consists of a 4400 kb closed circle, which is negatively supercoiled (13) and compacted about 1000-fold into a structure called the nucleoid within the bacterial cell (14). Although the nucleoid is not surrounded by a membrane, as in the eukaryotic nucleus, it appears to have some similarities in overall organization. Studies of the number of DNA nicks required to relax the chromosomal DNA completely have suggested the existence of around 50 discrete topological domains (15); this proposal is also supported by electron microscopy (14). A major difference between prokaryotic and eukaryotic DNA is that, while the writhing of DNA around histones accounts for virtually all the negative supercoiling in eukaryotic chromosomes, in pro-karyotes, some of the negative supercoiling is not accounted for by protein binding (16). Although a plasmid isolated from an $E.$ $coli$ cell has a typical σ value of -0.06, it has been estimated that approximately -0.025 of this is uncon-strained, that is, able to adopt the plectonemic conformation commonly seen in $vitro$ [see Section 6.1; (17)]; the remainder must consist of writhe and/or changes in twist stabilized by the binding of various kinds of proteins. It is assumed that the same proportions apply to the bacterial chromosome. This difference reflects the alternative methods for the introduction of negative super-coiling in the two types of cell. In eukaryotes, the formation of nucleosomes on initially relaxed DNA, followed by relaxation of the compensatory positive super-coiling by topoisomerases leads to negative supercoiling wholly constrained within nucleosomes (see Chapter 2, Section 5.5). In contrast, prokaryotes have an enzyme, DNA gyrase, which actively introduces negative supercoiling (see Chapter 5), as well as other topoisomerases capable of relaxing negative super-coils. Hence, the level of unconstrained supercoiling is controlled by the relative activities of competing topoisomerases and the presence of DNA-binding pro-teins which stabilize negative supercoiling.

Although a search for analogues of the eukaryotic histone proteins in bacteria has led to the discovery of a number of proteins which may be involved in the compaction of the DNA, structures as well defined as the nucleosome have not been discovered. The proteins involved in genome organization in prokaryotes are exemplified by the $E.$ $coli$ proteins HU and H1 (or H-NS) (18,19); both are responsible for compaction of the DNA. HU also wraps DNA in a negative sense (20,21) and has been shown to form relatively unstable nucleosome-like structures in electron micrographs (20). In addition to its function in genome organization, a specific gene regulation role has been proposed for protein H1 (22). RNA polymerase and nascent RNA chains have also been implicated in maintaining the structural integrity of the nucleoid (14).

3. Transcription: the twin supercoiled domain model

Consideration of the organization of chromosomes in both eukaryotes and prokaryotes seems to imply a reduction in the importance of supercoiling in the sense of free, torsionally-strained DNA. To summarize, it is suggested that most negative supercoiling in eukaryotes is constrained by binding to histones in nucleosomes and their higher-order aggregates, and that in prokaryotes (or *E. coli* at least), only some 40% of the supercoiling of isolated DNA is present *in vivo* as unconstrained plectonemic supercoiling (see Section 2). In recent years however, evidence has mounted that in bacterial cells, supercoiling is a dynamic process which depends on more than the balance between two opposing topoisomerases. In particular, the inactivation of topoisomerase I in *E. coli* by genetic means leads to high negative supercoiling of the plasmid pBR322, which is dependent on transcription from the strong promoter of the *tetA* (tetracycline resistance) gene on that plasmid (23,24). Furthermore, inhibition of DNA gyrase *in vivo* allows the isolation of positively supercoiled pBR322 (25).

In 1987, Liu and Wang (26) speculated that these observations might reflect a property of the transcription process itself. As RNA polymerase transcribes a gene, the protein–RNA complex must follow the helical path of the DNA strands. The most obvious view of this has the DNA stationary and the polymerase rotating around the DNA axis. However, Liu and Wang suggested that the combination of the polymerase, the nascent RNA chain, and possibly even ribosomes translating the mRNA in a co-ordinated fashion, would form a complex so large that it would be unable to rotate around the DNA (*Figure 6.2a*), and that instead, the DNA would rotate upon its axis. Such a rotation of the DNA around its axis, relative to the unpaired DNA region at the polymerase (*Figure 6.2b*), causes an increase in twist ahead of the moving polymerase, and a reduction in twist behind. Knowledge of the geometry of DNA supercoiling suggests that such twist changes would be manifested as positive and negative supercoiling respectively (*Figure 6.2c*). Because these constitute two regions of supercoiling separated by the polymerase complex, this has been termed the 'twin supercoiled domain' model. An alternative, but compatible view envisages that the polymerase itself is anchored to a large and essentially immobile matrix within the cell. The DNA is then translocated past the stationary polymerase, rotating about its axis as it does so, with the results described above; the evidence for this view has been reviewed (27).

Since the DNA strands are not broken during transcription, an overall change in the linking number of a DNA is impossible by this mechanism alone; the change in supercoiling ($Tw + Wr$) must be equal and opposite at each side of the polymerase. Of course, in a linear molecule, such transient supercoiling would diffuse away, and with a single transcription unit on a circular plasmid, the positive and negative supercoiling could cancel out by diffusion around the circle. However, as has been seen, DNA may be organized into discrete domains which are anchored to a large matrix, and which may be topologically independent. Such anchorage points would provide a block on the diffusion of transcription-

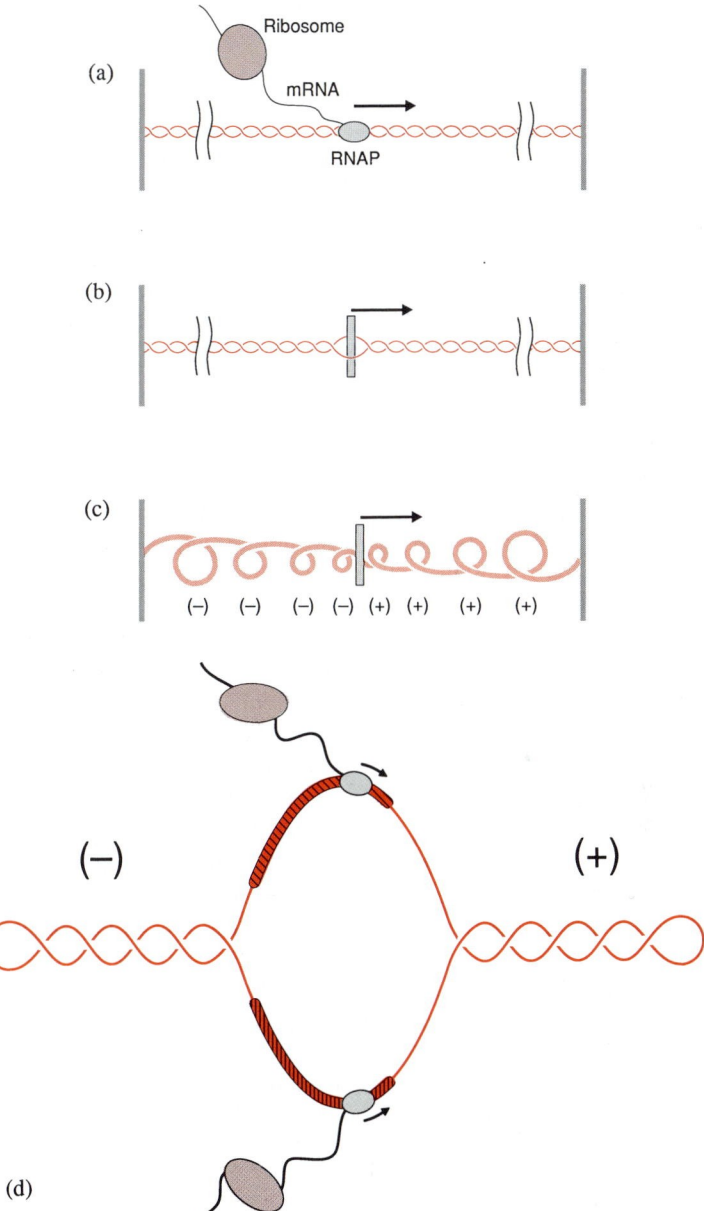

Figure 6.2. Formation of twin domains of supercoiling by transcription elongation. **(a)** RNA polymerase (RNAP) is shown transcribing a topologically isolated domain of DNA into mRNA, to which ribosomes become attached. This is shown schematically in **(b)** and leads to the formation of positive supercoiling ahead of the transcribing complex and negative supercoils behind **(c)**. **(d)** Domains of supercoiling can also occur in plasmid DNA with two transcription complexes travelling in opposite directions (redrawn from refs 26 and 28; the latter Copyright 1988 Cell Press).

generated supercoiling. Alternatively, transcription at a nearby gene proceeding in the opposite direction would also provide a blocking point (*Figure 6.2d*).

3.1. A new role for topoisomerases

This alternative mechanism for the generation of supercoiling in DNA suggests a new role for topoisomerases. In prokaryotes, topoisomerase I can only relax negative supercoils, whereas gyrase can efficiently relax positive supercoils (Chapter 5, Section 3). Hence it is likely that gyrase will act in front of, and topoisomerase I behind the transcription complex to remove the transiently formed supercoils. If the activities of the two topoisomerases are significantly different, as is the case in the inhibition experiments quoted above, this would result in the build-up of either negative or positive supercoils in a closed-circular plasmid.

The possible operation of this mechanism *in vivo* has been confirmed by experiments similar to those above, which demonstrate the requirement for transcription and the importance of the relative orientation of transcribing genes (28). The same effects have been observed in eukaryotic systems (29,30), where topoisomerases I and II can each relax both positive and negative super-coiling (see Chapter 5). The formation of twin supercoiled domains has even been demonstrated *in vitro* using transcription from a single promoter on a closed-circular plasmid in the presence of *E. coli* topoisomerase I (31). Transcription-dependent increases in negative supercoiling have also been observed without inhibition of topoisomerases, suggesting that such effects can occur in wild-type cells (32). The magnitude of these effects may be very great, particularly in the vicinity of actively transcribed genes, and it has been sug-gested that transiently high levels of supercoiling may have a role in regulatory processes. The idea, for example, that altered DNA conformations such as Z-DNA and cruciforms (see Chapter 1, Sections 2.4 and 3.1) may be important *in vivo* now seems more acceptable. Earlier estimations of the overall average levels of unconstrained supercoiling in cells were too low to be consistent with the formation of such structures (see Section 2.2). Support for the existence of Z-DNA in *E. coli* has come from the observation that a recognition site is not a substrate for its specific methylase when the site is within a sequence with Z-DNA-forming potential. The same site is a substrate when present in right-handed B-form DNA (33). Subsequently, similar experiments have suggested a role for transcription-induced supercoiling in the formation of Z-DNA *in vivo* (34).

4. Control of gene expression

Having discussed the process of transcription elongation, or the movement of RNA polymerase along the DNA template, and its consequences in terms of the supercoiling of the template, the more general effects of DNA topology on the control of gene expression will now be addressed. Gene expression in its broadest sense encompasses all the factors which determine whether protein is produced from a given gene, or group of genes. DNA topology, specifically DNA supercoiling, is an important factor influencing gene expression, at the level of transcription initiation. The initiation of transcription involves the binding of RNA polymerase to the promoter region of a gene, and its progression to continuous RNA synthesis through a number of discrete steps. DNA supercoiling affects this process both directly, and also indirectly, through its influence on the binding of repressor and activator proteins to regions surrounding the site of transcription initiation. These two aspects will be discussed separately.

4.1. RNA polymerase binding

The processes involved in RNA polymerase binding will be discussed in terms of prokaryotes, as exemplified by *E. coli* (35). It is to be expected that the general principles, if not the specifics, will be applicable to eukaryotic transcription, about which there is much less detailed information (36).

Most promoter sequences in *E. coli* contain variations on the '-10' and '-35' consensus regions (the numbers refer to positions upstream of the transcription start point), which are recognized by the complex of RNA polymerase and the sigma factor, σ^{70}, responsible for the specificity of binding. The initial RNA polymerase complex is known as the 'closed' complex, and, as has been mentioned (Chapter 2, Section 5.5), it involves the wrapping of the DNA around the protein. The linking difference (ΔLk) associated with the closed complex is around -1.25 (37), that is, the wrapping is left-handed. The closed complex is then converted to the 'open' complex; the ΔLk is now about -1.6 (37,38), and the strands around the -10 region become completely unwound and separated over a length of 12 bp (39,40). It is hypothesized that the negative writhe associated with the closed complex is converted to untwisting of the promoter in the open complex (41,42); in addition to the strand separation, the open complex must also contain some residual wrapping or untwisting. The open complex is able to begin synthesis of RNA; however, after the addition of a few nucleotides, the enzyme may release the transcript and begin synthesis again, in a process known as abortive cycling. Transcription proper begins when the σ factor dissociates from the complex and RNA synthesis proceeds beyond the promoter region. The steps involved in transcription initiation are outlined in *Figure 6.3*.

It is to be expected that both the binding of RNA polymerase to form the closed complex, and the isomerization to the open complex will be promoted by increasing negative supercoiling. The closed complex stabilizes the writhing associated with negative supercoiling, and the under-winding of the DNA helix caused by supercoiling will promote the strand separation in the open complex (43). This suggests that transcription initiation is likely to be promoted by increasing levels of supercoiling, and indeed it was shown in the 1970s that in general, levels of transcription are increased on negatively supercoiled templates [e.g. (44)]. The response of individual promoters to supercoiling is more complex; of those that have been tested, a majority are stimulated by increasing negative supercoiling, some are unaffected, and a few are inhibited (45). The

(a)

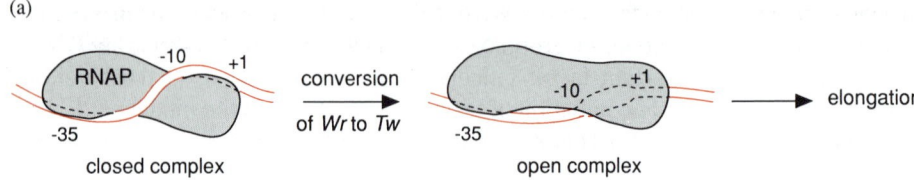

closed complex   open complex

(b)

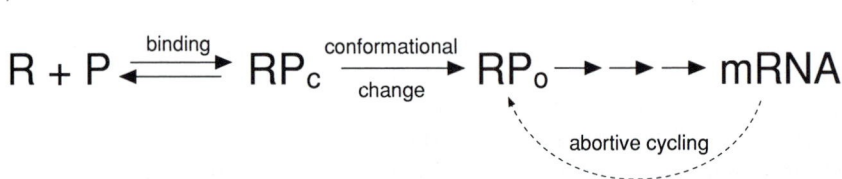

Figure 6.3. Transcription initiation at *E. coli* promoters. **(a)** The initial closed complex between RNA polymerase (RNAP) and the promoter involves the winding of DNA around the protein in a left-handed sense. The numbers indicate the approximate position of the -35 and -10 regions and the transcription start point ($+1$). Isomerization to the open complex results in the conversion of writhe to twist, and the unpairing of a section of helix around the -10 region, prior to RNA synthesis. **(b)** The same process, shown symbolically. R, RNA polymerase; P, promoter; RP_c, closed complex; RP_o, open complex. The abortive cycling step, after the synthesis of a short oligoribonucleotide is also indicated (see text).

precise effect is likely to be modulated by the sequence and local conformation of the -35 and -10 regions, as well as the spacing between them (46).

The variation of response to supercoiling in different promoters is exemplified by the genes encoding *E. coli* topoisomerase I and DNA gyrase. It is thought that the transcription of these genes is oppositely regulated in a homeostatic mechanism to maintain an appropriate level of supercoiling *in vivo*. Increased negative supercoiling increases the initiation of transcription from the topoisomerase I gene (*topA*) (47) and reduces initiation from the *gyrA* and *gyrB* genes encoding the subunits of DNA gyrase (48), contrary to the expected effect of supercoiling on DNA binding and unwinding by RNA polymerase. In the case of the gyrase genes, this effect has been shown to be dependent on the DNA sequence surrounding the site of initiation of RNA synthesis (the $+1$ position), and it has been proposed that this may be mediated by an effect on the stability of the abortive cycling complex; relaxation of the template may cause destabilization of the complex and progression to elongation (49).

It has also been suggested that the intracellular level of supercoiling may be used as a co-ordinate regulator of a number of genes by environmental conditions. For example, changes in anaerobic versus aerobic growth and changes in the osmolarity of the medium influence the intracellular levels of supercoiling. One possibility is that this effect is mediated through the intracellular [ATP]/

[ADP] ratio and its effect on the activity of DNA gyrase (50,51). This mechanism has been implicated in the regulation of a number of genes (52). Finally it should be noted that at very high levels of negative supercoiling, beyond those normally present *in vivo*, overall initiation of transcription is inhibited. A possible explanation for this is the formation of alternative DNA structures such as cruciforms or Z-DNA, which may not be efficient templates for RNA polymerase (see Chapter 1).

4.2. Transcriptional regulatory proteins

Another way in which DNA topology may influence the control of gene expression relates to the binding of regulatory proteins which affect (activate or repress) transcription by RNA polymerase. Many such proteins are now known, in both prokaryotes and eukaryotes, and some have been subjected to high-resolution structure determination by X-ray crystallography or NMR (53,54). The binding of such proteins to their DNA target sites can be influenced by DNA topology, and the formation of the protein–DNA complex can have topological consequences.

The *ara* operon of *E. coli* consists of three genes involved in the utilization of arabinose, transcribed from a single promoter. This promoter is controlled (both positively and negatively) by the co-operative binding of the AraC protein to two operator sites some 200 bp apart. Although the binding to one of the operators is tight, and to the other is weak, the two sites are normally occupied by the AraC protein (55). Moreover, insertion and deletion of small segments of DNA in the region between the two operators leads to cyclical changes in the repression of the *ara* promoter by AraC, with a period consistent with the helical repeat of DNA (56). These findings and other data are consistent with the co-operative binding of AraC at the two operators via 'looping' of the intervening DNA. Other experiments have shown that loop formation by AraC can be promoted by DNA supercoiling. This is presumably due to the greater ease with which remote sites in plectonemically supercoiled DNA can be brought together.

One of the best known and most studied of the prokaryotic repressor proteins is that from bacteriophage λ, usually called λ repressor or cI protein (57). This protein binds as a dimer to three operator sites which overlap promoters important in determining whether the phage will enter the lytic (replicative) or lysogenic (quiescent) phase of growth. The binding of repressor dimers to the three operator sites is co-operative and, in *in vitro* constructs, can also apparently occur by looping of the DNA between the operator sites (58). In experiments by Hochschild and Ptashne, operator binding sites were separated by varying distances and the degree of co-operativity assessed. When the operator sites were separated by an integral member of helical turns (e.g. 5 or 6), co-operativity was observed, whereas when they were separated by a non-integral number (e.g. 4.6, 5.5, or 6.4) co-operativity was lost. This was interpreted as suggesting that two repressor dimers must be bound on the same side of the DNA helix to come together by smooth bending of the intervening DNA (*Figure 6.4*). Sites separated by non-integral turns cannot interact in this way unless the intervening DNA is twisted or writhed, which is likely to be thermodynamically less favourable than smooth bending (*Figure 6.4*).

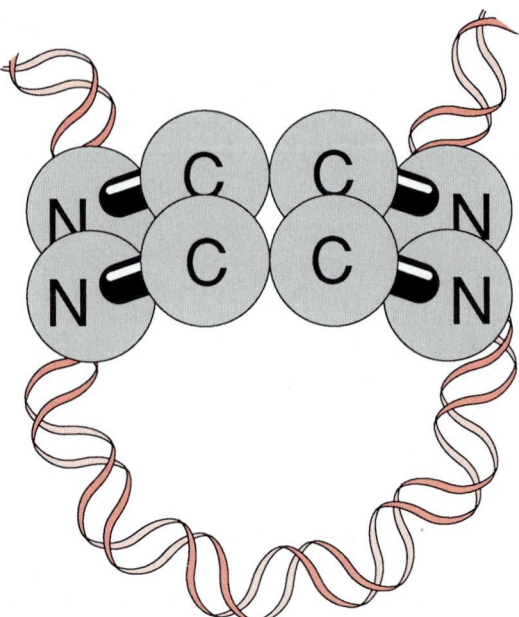

Figure 6.4. Looping of DNA. Co-operative interaction of a pair of bacteriophage λ repressor dimers via their C-terminal domains is proposed to occur by the smooth deformation (looping) of the DNA between the operator sites (redrawn by permission from ref. 88. Copyright © 1986 Macmillan Magazines Limited).

Evidence for DNA looping has also been found in other systems. The *lac* repressor is a tetrameric protein that binds to the operator site of the *lac* operon and prevents transcription from the adjacent promoter site (53). Using a gel electrophoretic method, Müller-Hill and co-workers have shown that a single *lac* repressor tetramer can bind to two operator sites separated by several hundred nucleotides, provided the sites are on the same side of the DNA (59). The interpretation is that this permits the formation of a DNA loop anchored by the repressor protein.

A loop in DNA stabilized by protein–protein interactions constitutes a discrete topological domain and, up to a point, can be treated in much the same way as a closed DNA circle (see Chapter 2). It is likely that the primary structural features of DNA that influence DNA curvature and flexibility (see Chapter 1, Section 4) will also influence the ability of DNA to form loops.

Another feature of DNA–protein interactions is the change in twist which may be imparted to the DNA upon protein binding. Section 4.2 of Chapter 1 discussed the binding of bacteriophage 434 repressor protein to DNA, which may in part be determined by the ease of flexure of the central four base pairs of the operator sequence. Other experiments support the notion that the repressor alters the twist of the DNA and that the intrinsic twist of the operator determines repressor affinity (60). Another system where twist changes in DNA have been found to be important is in the mercury genes of *E. coli*. Here, RNA polymerase binds to the promoter at which the MerR repressor is bound, but cannot proceed to transcribe until Hg(II) ion binds to the repressor. Experiments suggest that the MerR–Hg(II) complex distorts the DNA in the promoter region to induce under-winding (61). The effect of this distortion is proposed to lead to the optimal alignment of DNA elements in the promoter region important for RNA polymerase binding (see Section 4.1). Clearly, protein binding which leads to alterations in DNA twist will be profoundly affected by the supercoiled state of the DNA.

For a number of DNA-binding proteins, high resolution structural information is available for the protein complexed to its DNA target site (53,54). Several of these show marked distortions of the DNA, in particular bending of the DNA helix. Perhaps the most spectacular of these is the structure of CAP (see Chapter 1, Section 4.2); other examples include the bacteriophage 434 repressor previously described and the *Eco*RI restriction enzyme (62).

5. Replication

Since the realization that the *E. coli* chromosome is a single closed circle, the topological problem associated with the unwinding of the DNA helix during replication has been recognized (63). This turns out to be analogous to the topological consequences of transcription elongation discussed in Section 3. The topological state of the DNA also influences the mechanisms of initiation and termination of replication; these aspects will be discussed separately.

5.1. Replication initiation

In all systems, replication of DNA proceeds from specific sites known as 'origins of replication'. In the majority of cases, DNA synthesis is preceded by the binding of sequence-specific proteins at the origin. It has been shown that these binding events are often stimulated by negative supercoiling of the origin. For example, the complex of the initiation protein, DnaA and *oriC*, the origin of replication in the *E. coli* chromosome, is more stable if *oriC* is present on a negatively supercoiled DNA (64). This is analogous to the stimulation of binding of RNA polymerase (Section 4), and implies a wrapping of the DNA in a negative sense in the initiation complex. A fundamental step in the initiation of replication is of course the unwinding and unpairing of a region of the DNA helix. In the case of DnaA protein, unwinding of an AT-rich region of DNA adjacent to the origin quickly follows binding, but is absolutely dependent on negative supercoiling of the DNA (65). A similar requirement has been demonstrated for the bacteriophage λ O protein, which binds to the replication origin of bacteriophage λ (66). A further demonstration of the importance of supercoiling in replication initiation is provided by the yeast autonomously replicating sequences (ARS), which unwind readily when present in negatively supercoiled DNA. Their ease of unwinding is correlated with their efficiency as replication origins *in vivo* (67).

In bacteria, DNA gyrase has been shown to be required for the initiation of replication, but this requirement is thought to relate to its supercoiling activity, rather than to a specific role of the protein at the origin.

5.2. Replication elongation

As the replication fork proceeds along a DNA template, the unwinding of the strands causes positive supercoils to build up ahead. This effect is analogous to the situation in transcription (Section 3), except that no corresponding domain of negative supercoiling occurs, since the parent strands are not paired again. In the case of *E. coli*, DNA gyrase is the obvious choice of topoisomerase to remove the positive supercoils, which would otherwise rapidly inhibit elongation, since it is efficient at relaxation of positive supercoils, whereas topoisomerase I will only relax negative supercoils (see Chapter 5).

In the yeasts *Saccharomyces cerevisiae* and *Schizosaccharomyces pombe*, the elongation of DNA synthesis is much reduced by inactivation of both topoisomerases I and II, but not by the inactivation of either (68), implying that in eukaryotes, either topoisomerase I or II can act to relieve the strain caused by elongation; this is to be expected since both enzymes can efficiently remove positive supercoils (see Chapter 5, *Table 5.1*).

5.3. Termination of replication

When two replication forks converge at the end of DNA synthesis during replication, the unwinding of the parental DNA strands may not coincide with their replication to two daughter DNA helices (see *Figure 6.5*). The product of replication may then be a pair of catenated DNA rings, or catenated topological domains in the case of linear replicons. Such structures have been documented as intermediates in replication in a number of systems; as discussed in Chapter 4, Section 3.1 (69).

In principle, catenated DNA molecules may be resolved by either type I or type II topoisomerases, but in the case of a type I, the reaction requires a nick, or single-stranded gap in at least one of the substrate DNAs (see Chapter 5). Therefore, if the component rings of the product catenane are converted to a closed double-stranded form (*Figure 6.5c*), only a type II topoisomerase would be able to unlink the products. In *E. coli*, DNA gyrase has been shown to decatenate catenanes formed by bacteriophage λ Int protein [see Section 6.1; (17)], and gyrase has been shown to be essential for proper segregation of the chromosome, which is consistent with a role in the decatenation of daughter chromosomes (70). In yeast, catenanes of a closed-circular plasmid have been shown to accumulate on inactivation of topoisomerse II (71), and it has been shown that the enzyme is required during mitosis, again implying a role in chromosome segregation (72).

The results above provide strong evidence for the involvement of type II enzymes in the decatenation of daughter replicons; however, it has been sug-

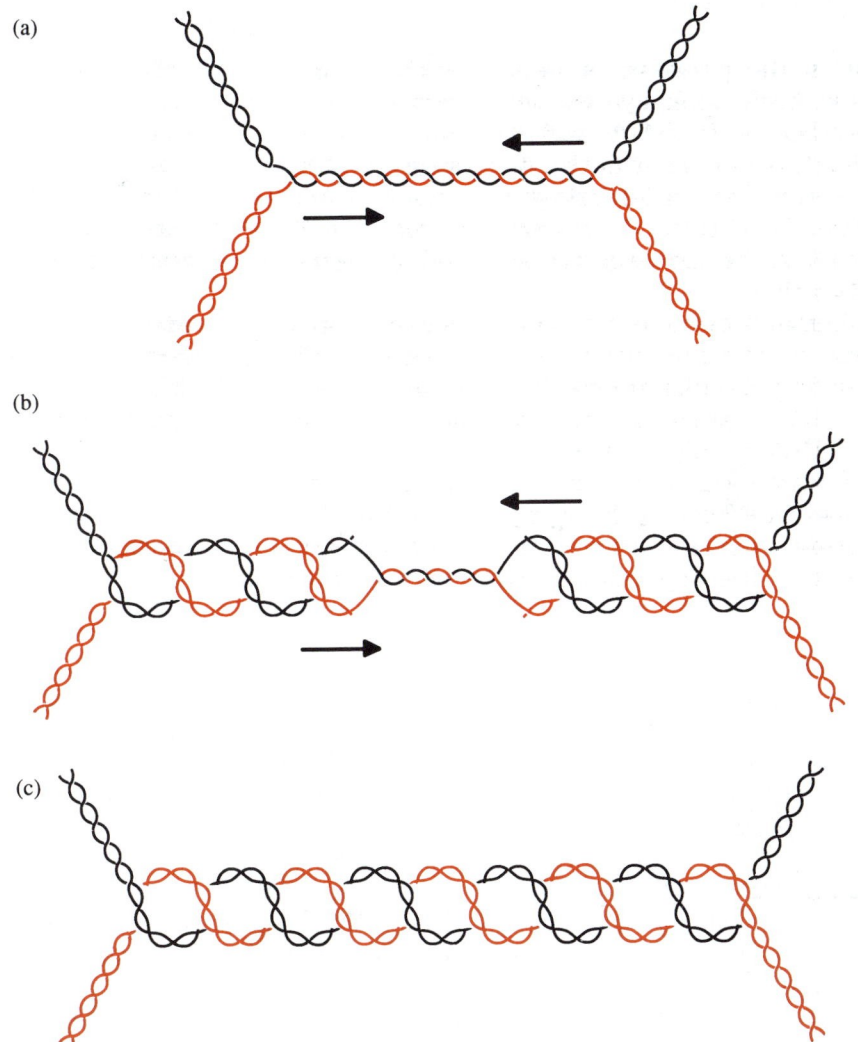

Figure 6.5. Formation of catenated DNA at the termination of replication. At the terminus of replication, converging replication forks, **(a)**, lead to the interwinding of daughter molecules, **(b)**. Upon completion of replication, the products are catenated DNA circles, **(c)**.

gested that type I topoisomerases may also be involved. For example, in the replication of pBR322 in a purified system, it was shown that *E. coli* topoisomerase I could decatenate the product molecules, if present in a high enough concentration (73). Presumably, the high concentration ensures that decatenation occurs before the molecules become fully closed-circular.

6. Recombination

Recombination is the process of genetic exchange in DNA which rearranges chromosomes in both prokaryotes and eukaryotes. It involves the breaking of phosphodiester bonds in DNA and the rejoining of the broken ends to new partners, as illustrated earlier in the discussion of Holliday junctions (Chapter 1, Section 3.1). Generalized or homologous recombination is abundant in all species, and is exemplified by the RecA protein of *E. coli*. The RecA protein is known to unwind duplex DNA; hence negative supercoiling is expected to stimulate homologous recombination.

A large number of recombination reactions in both prokaryotes and eukaryotes are now known to involve interaction between specific DNA sequences (so-called site-specific recombination). The specific sequences are aligned and the exchange of DNA strands is brought about by recombination proteins. The importance of DNA topology in these processes cannot be over-emphasized, both in terms of facilitating the recombination reaction and in terms of understanding the recombination mechanism by analysis of the topology of the products. The following sections describe two examples of well-studied recombination reactions, highlighting the aspects relevant to DNA topology.

6.1. Bacteriophage λ Int protein

As illustrated in *Figure 6.6*, the bacteriophage λ Int protein catalyses a recombination reaction between a site on the bacteriophage DNA (*att*P) and a site on the host DNA (*att*B) which leads to the integration of λ DNA into the *E. coli* chromosome (74). This reaction has been extensively studied *in vitro*, generally using artificial DNA substrates such as those shown in *Figure 6.6b*. DNA supercoiling has been shown to have a profound effect on the λ integration reaction. Only negatively supercoiled closed-circular DNA molecules bearing the *att*P site are effective substrates for recombination. Supercoiling at *att*B is not necessary. The requirement for supercoiling of the bacteriophage DNA is thought to relate to the folding of *att*P into a higher-order structure in which the DNA is wrapped on itself to create a negative writhe (75). This higher-order structure has been termed the 'intasome', by analogy with the nucleosome (see Chapter 3, Section 5), and comprises *att*P complexed with Int protein and IHF (integration host factor). This DNA–protein complex constitutes a binding site for *att*B which probably binds to the complex as a naked piece of DNA.

DNA substrates which carry both the *att*P and *att*B sites (e.g. *Figure 6.6b*) have been shown to generate knotted or catenated products, depending on the relative orientation of the *att*P and *att*B sites. Double-stranded circular DNA substrates containing *att*P and *att*B in the same orientation generate catenanes (76,77) while those containing *att* sites in an inverted orientation generate knots (76,78). Analysis of the knotted and catenated products has suggested likely models for the bacteriophage λ integration reaction. For example, the complexity of the knots formed by recombination increases with increasing specific

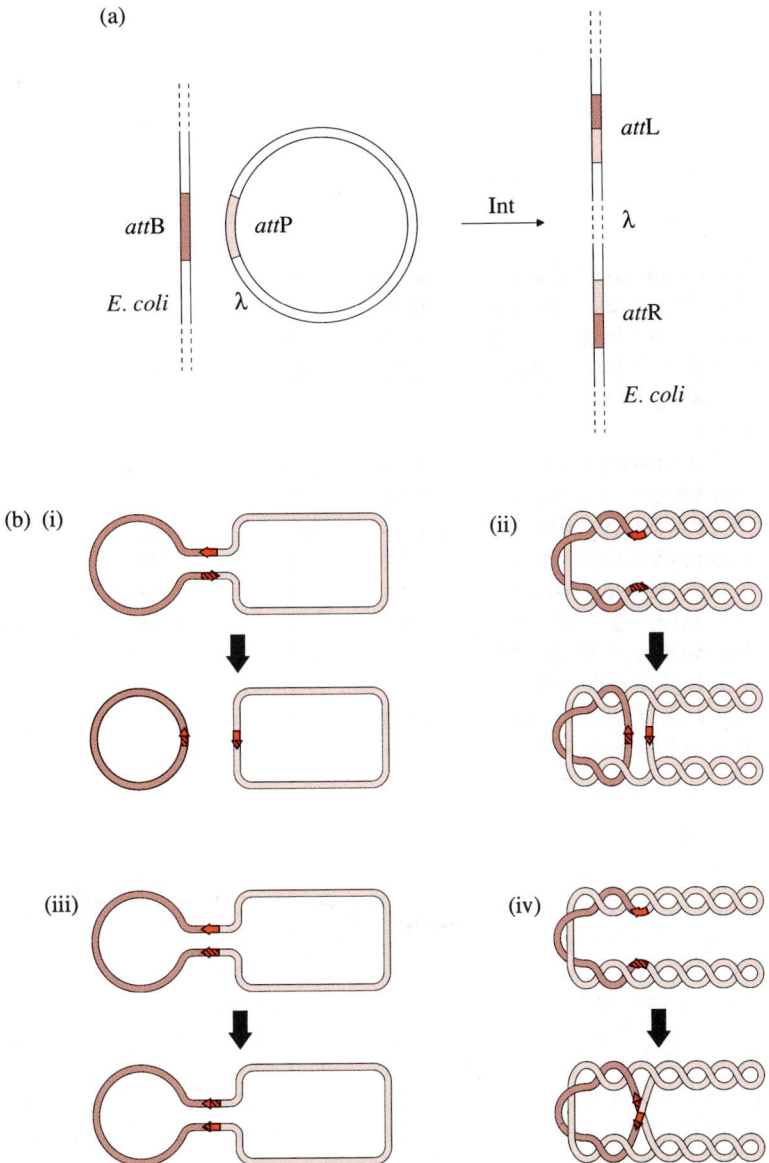

Figure 6.6. Site-specific recombination by the bacteriophage λ Int protein. **(a)** Bacteriophage λ DNA is integrated into the *E. coli* chromosome via a recombination event between the *att*P site on λ DNA and the *att*B site on *E. coli* DNA, mediated by Int and other proteins. The integrated phage (the prophage) is now flanked by two hybrid sites, *att*L and *att*R. **(b)** Recombination between *att* sites (shown as arrows) in direct repeat on the same circle generates two product circles, (i) which, if the substrate DNA was plectonemically supercoiled, will be catenated, (ii). When *att* sites are in inverted repeat (iii), the product is a single circle which will be knotted if the substrate was plectonemically supercoiled (iv) (b redrawn from ref. 78).

linking difference of the substrate. This is due to the trapping of writhes (or nodes) in interwound DNA (see Chapter 2, Section 4.2) between the interacting recombination sites. Non-supercoiled (nicked-circular) DNA molecules can, under certain conditions, constitute substrates for recombination and, at low frequency, generate both simple (unknotted) and knotted products. Increasing the distance between the *att* sites in supercoiled DNA substrates also increases the complexity of the knots produced. It has been concluded from these (and other) data that synapsis between *att* sites occurs by random collision, which leads to the trapping of supercoils (writhes) between the two *att* sites which will then be converted to knot or catenane nodes following recombination (*Figure 6.6b*). The knotted products of recombination between *att* sites on non-supercoiled DNA can be explained by formation of the intasome which may lead to the trapping of a writhe in the DNA between the *att* sites and hence knotted products.

The bacteriophage λ integration reaction can also be used as a probe of the conformation of supercoiled DNA *in vivo*. For example, it has been concluded that the writhing of DNA in *E. coli* is plectonemic rather than toroidal. Toroidal supercoiling would not be expected to generate knotted and catenated products of the complexity found in the bacteriophage λ integration reaction [see Chapter 2, Section 4.2.1; (77,79)]. Furthermore, analysis of the catenated products of Int-mediated recombination *in vivo*, has concluded that only 40% of the linking deficit of plasmid DNA in *E. coli* is in the form of interwound (plectonemic) supercoils (17). The remainder is thought to be constrained by, for example, protein binding (see Section 2.2).

6.2. Resolvase

Another well-studied site-specific recombination system is the reaction catalysed by the resolvase proteins from the transposons of the Tn3 family. Transposons are segments of DNA that can apparently 'jump' from one molecule to another (8). In the case of Tn3 this process occurs in two stages. The first involves the fusion of the donor and recipient DNA molecules to yield a co-integrate bearing two copies of the transposon, one at each junction of the donor and recipient DNA (*Figure 6.7a*). This co-integrate is now resolved into the product DNA molecules, each bearing a single copy of the transposon, by the action of the resolvase protein which catalyses a site-specific recombination reaction between the *res* sites in each transposon. The product DNA molecules are catenated (*Figure 6.7a*), and may subsequently be decatenated by the action of a topo-isomerase such as DNA gyrase.

The site-specific recombination reaction catalysed by resolvase can be studied *in vitro* using DNA molecules carrying pairs of *res* sites (81). Such studies have shown that, unlike bacteriophage λ Int, resolvase will only efficiently recombine substrates bearing directly repeated *res* sites (82) (*Figure 6.7*). Moreover, the major product of the resolvase recombination reaction is a singly-linked catenane, regardless of the specific linking difference of the DNA substrate (*Figure 6.7b*).

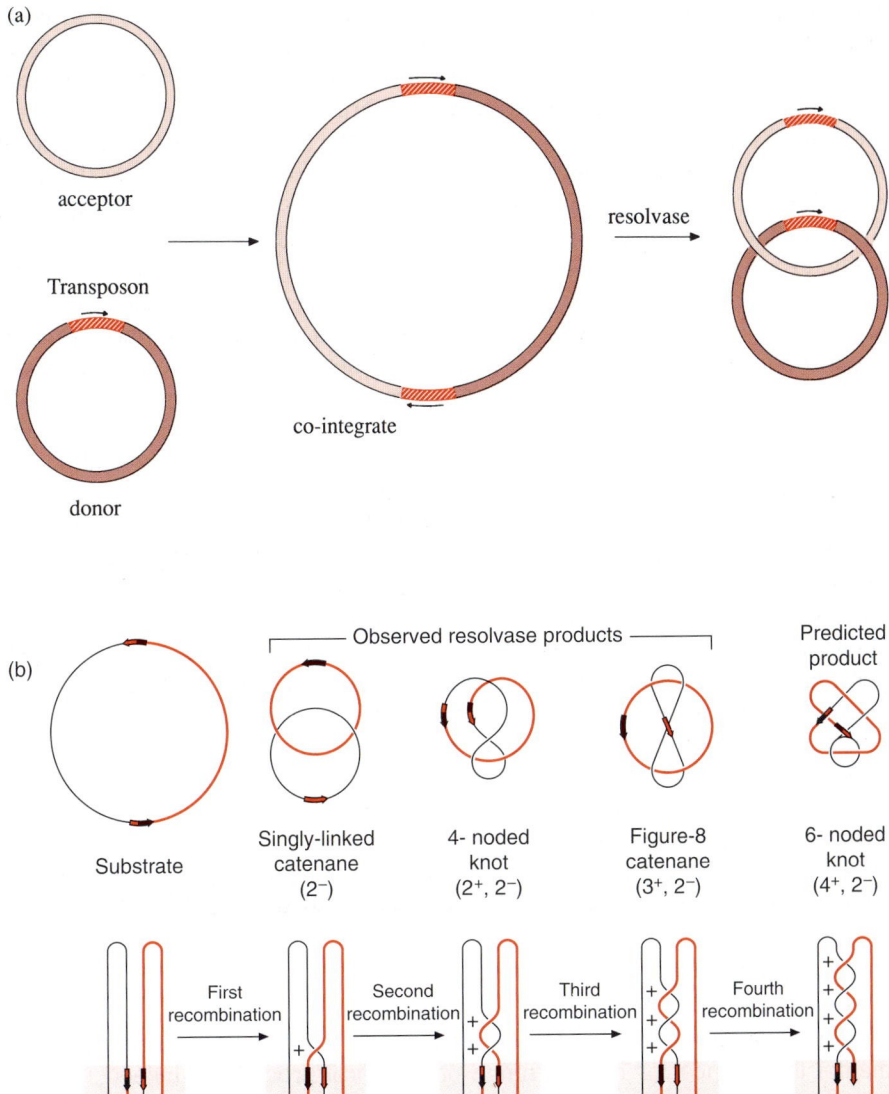

Figure 6.7. Site-specific recombination by resolvase. **(a)** A donor circle carrying a transposon of the Tn3 family forms a co-integrate with the acceptor DNA circle. The co-integrate carries two copies of the transposon. The action of resolvase catalyses a recombination reaction between *res* sites within the two transposons to yield the catenated product circles, each carrying a single copy of the transposon. **(b)** Scheme for the formation of multiple knotted and catenated products by Tn3 resolvase. Successive rounds of recombination generate the products shown in the upper row. These products can be rationalized by proposing a synaptic process involving three nodes (lower row) (b redrawn from ref. 84. Copyright 1985 by the AAAS).

The simplicity of this product implies a rigidly controlled mechanism of synapsis and not simply random collision as has been suggested for the bacteriophage λ Int protein. Careful analysis of the products of resolvase-mediated recombination has revealed that, in addition, a range of knotted and catenated products are present at low frequency (83,84) (*Figure 6.7b*). These products can arise by successive rounds of recombination. It is important to note that only the products with the node assignation given in *Figure 6.7b* were detected, again suggesting a rigidly controlled synaptic process (e.g. there are eight possible 6-noded knots, but only the one shown in *Figure 6.7b* has been detected). It has been proposed that the resolvase-mediated synapsis of *res* sites involves the stabilization of three negative supercoils (i.e. three negatives nodes) as shown in the lower panel of *Figure 6.7b*. Indeed, measuring ΔLk for the synaptic complex involving resolvase and the two *res* sites gives a value of -3 consistent with this proposal (85).

The requirement for directly repeated *res* sites has prompted much speculation as to the mechanism of bringing together the two sites. One model, known as tracking, involves the binding of the resolvase protein to one *res* site via one domain, and another domain of the protein sliding along the DNA until the other *res* site is encountered. Several lines of experimentation now argue against this type of mechanism (reviewed in ref. 3). Perhaps the most convincing of these is the construction of catenanes containing the two recombination sites on separate DNA circles. Provided the correct orientation of sites is maintained, recombination can still occur. Such a result is difficult to reconcile with a tracking model. A more likely explanation of the orientation requirement is that the correct synaptic structure can only be assembled on plectonemically wound negatively supercoiled DNA with the *res* sites in directly repeated orientation (86,87). The process of slithering of the DNA (see Chapter 4, Section 4) can be invoked to align the *res* sites correctly. Thus the requirement for directly repeated sites arises from the negatively supercoiled state of the DNA. Indeed, it has been shown that orientation is unimportant with linear substrates (86).

7. Conclusions

The examples given in this chapter demonstrate that DNA topology is of fundamental importance for a wide range of biological processes. Indeed, virtually every reaction involving DNA is influenced by DNA topology, or has topological effects. This book has progressed from the somewhat abstract concepts of linking number, twist, writhe, etc. through to their ultimate biological consequences. It is hoped that an appreciation of some of the finer points of DNA topology will provide greater insight into the biological functions of DNA.

8. Further reading

Bellomy,G.R. and Record,M.T. (1990) Stable DNA loops *in vivo* and *in vitro*: roles in gene regulation at a distance and in biophysical characterization of DNA. *Prog. Nucleic Acids Res. Mol. Biol.*, **39**, 81.

Drlica, K. (1984) Biology of bacterial deoxyribonucleic acid topoisomerases. *Microbiol. Rev.*, **48**, 273.

Drlica, K. (1987) The nucleoid. In Neidhardt, F. C. (ed.), *Escherichia coli and Salmonella typhimurium*. ASM Press, Washington DC, p. 91.

Horwitz, M. S. Z. and Loeb, L. A. (1990) Structure–function relationships in *Escherichia coli* promoter DNA. *Prog. Nucleic Acids Res. Mol. Biol.*, **38**, 137.

Nash, H. A. (1990) Bending and supercoiling of DNA at the attachment site of bacteriophage λ. *Trends Biochem. Sci.*, **13**, 222.

Pederson, D. S., Thoma, F., and Simpson, R. T. (1986) Core particle, fiber, and transcriptionally active chromatin structure. *Annu. Rev. Cell. Biol.*, **2**, 117.

Ptashne, M. (1986) Gene regulation by proteins acting nearby and at a distance. *Nature*, **322**, 697.

Stark, W. M., Boocock, M. R., and Sherratt, D. J. (1989) Site-specific recombination by Tn3 resolvase. *Trends Genet.*, **5**, 304.

Wang, J. C. and Giaever, G. N. (1988). Action at a distance along a DNA. *Science*, **240**, 300.

Wang, J. C. and Liu, L. F. (1990) DNA replication: topological aspects and the roles of DNA topoisomerases. In Cozzarelli, N. R. and Wang, J. C. (ed.), *DNA topology and its biological effects*. Cold Spring Harbor Laboratory Press, Cold Spring Harbor, p.321.

9. References

1. Gasser, S. M. and Laemmli, U. K. (1987) *Trends Genet.*, **3**, 16.
2. Boles, T. C., White, J. H., and Cozzarelli, N. R. (1990) *J. Mol. Biol.*, **213**, 931.
3. Gellert, M. and Nash, H. (1987) *Nature*, **325**, 401.
4. Parker, C. N. and Halford, S. E. (1991) *Cell*, **66**, 781.
5. Lilley, D. M. J. and Hallam, L. R. (1984) *J. Mol. Biol.*, **180**, 179.
6. Svaren, J. and Chalkley, R. (1990) *Trends Genet.*, **6**, 52.
7. Brown, D. D. (1984) *Cell*, **37**, 359.
8. Felsenfeld, G. (1992) *Nature*, **355**, 219.
9. Morse, R. H. (1992) *Trends Biochem. Sci.*, **17** 23.
10. Thoma, F., Koller, T. and Klug, A. (1979) *J. Cell. Biol.*, **83**, 403.
11. Widom, J. (1989) *Annu. Rev. Biophys. Biophys. Chem.*, **18**, 365.
12. Thomas, J. O. (1984) *J. Cell. Sci. Suppl.*, **1**, 1.
13. Worcel, A. and Burgi, E. (1972) *J. Mol. Biol.*, **71**, 127.
14. Drlica, K. (1987) In Neidhardt, F. C. (ed.), *Escherichia coli and Salmonella typhimurium*. ASM Press, Washington DC, p.91.
15. Sinden, R. R. and Pettijohn, D. E. (1981) *Proc. Natl. Acad. Sci. USA*, **78**, 224.
16. Sinden, R. R., Carlson, J., and Pettijohn, D. E. (1980) *Cell*, **21**, 773.
17. Bliska, J. B. and Cozzarelli, N. R. (1987) *J. Mol. Biol.*, **194**, 205.
18. Schmid, M. B. (1990) *Cell*, **63**, 451.
19. Rouvière-Yaniv, J., Kiseleva, E., Bensaid, A., Almeida, A., and Drlica, K. (1991) In Mohan, S. B., Dow, C. and Cole, J. A. (ed.), *Prokaryote structure and function: a new perspective. Society for General Microbiology symposium*. Cambridge University Press, Cambridge, p.414.
20. Rouvière-Yaniv, J., Yaniv, M., and Germond, J.-E. (1979) *Cell*, **17**, 265.
21. Broyles, S. S. and Pettijohn, D. E. (1986) *J. Mol. Biol.*, **187**, 47.
22. Hulton, C. S. J., Seirafi, A., Hinton, J. C. D., Sidebotham, J. M. Waddell, L., Pavitt, G. D., Owen-Hughes, T., Spassky, A., Buc, H., and Higgins, C. F. (1990) *Cell*, **63**, 631.
23. Pruss, G. J. (1985) *J. Mol. Biol.*, **185**, 51.
24. Pruss, G. J. and Drlica, K. (1986) *Proc. Natl. Acad. Sci. USA*, **83**, 8952.
25. Lockshon, D. and Morris, D. R. (1983) *Nucleic Acids Res.*, **11**, 2999.

26. Liu,L.F. and Wang,J.C. (1987) *Proc. Natl. Acad. Sci. USA*, **84**, 7024.
27. Cook,P.R. (1989) *Eur. J. Biochem.*, **185**, 487.
28. Wu,H.-Y., Shyy,S., Wang,J.C., and Liu,L.F. (1988) *Cell*, **53**, 433.
29. Giaever,G.N. and Wang,J.C. (1988) *Cell*, **55**, 849.
30. Brill,S.J. and Sternglanz,R. (1988) *Cell*, **54**, 403.
31. Tsao,Y.-P., Wu, H.-Y., and Liu,L.F. (1989) *Cell*, **56**, 111.
32. Figueroa,N. and Bossi,L. (1988) *Proc. Natl. Acad. Sci. USA*, **85**, 9416.
33. Jaworski,A., Hsieh,W.T., Blaho,J.A., Larson,J.E., and Wells,R.D. (1987) *Science*, **238**, 773.
34. Rahmouni,A.R. and Wells,R.D. (1992) *J. Mol. Biol.*, **223**, 131.
35. McClure,W.R. (1985) *Annu. Rev. Biochem.*, **54**, 171.
36. Geiduschek,E.P. and Tocchini-Valentini,G.P. (1988) *Annu. Rev. Biochem.*, **57**, 873.
37. Amouyal,M. and Buc,H. (1987) *J. Mol. Biol.*, **195**, 795.
38. Gamper,H.B. and Hearst,J.E. (1982) *Cell*, **29**, 81.
39. Kirkegaard,K., Buc,H., Spassky,A., and Wang,J.C. (1983) *Proc. Natl. Acad. Sci. USA*, **80**, 2544.
40. Spassky,A., Kirkegaard, K., and Buc,H. (1985) *Biochemistry*, **24**, 2723.
41. Buc,H. (1986) *Biochem. Soc. Trans.*, **14**, 196.
42. Travers,A.A. (1986) *Biochem. Soc. Trans.*, **14**, 199.
43. Drew,H.R., Weeks,J.R., and Travers,A.A. (1985) *EMBO J.*, **4**, 1025.
44. Hayashi,Y. and Hayashi,M. (1971) *Biochemistry*, **10**, 4212.
45. Drlica,K. (1984) *Microbiol. Rev.*, **48**, 273.
46. Borowiec,J.A. and Gralla,J.D. (1987) *J. Mol. Biol.*, **195**, 89.
47. Tse-Dinh,Y.-C. (1985) *Nucleic Acids Res.*, **13**, 4751.
48. Menzel,R. and Gellert,M. (1983) *Cell*, **34**, 105.
49. Menzel,R. and Gellert,M. (1987) *Proc. Natl. Acad. Sci. USA*, **84**, 4185.
50. Hsieh,L.-S., Rouvière-Yaniv,J., and Drlica,K. (1991) *J. Bact.*, **173**, 3914.
51. Hsieh,L.-S., Burger,R.M., and Drlica,K. (1991) *J. Mol. Biol.*, **219**, 443.
52. Higgins,C.F., Dorman,C.J., Stirling,D.A., Waddell,L., Booth,I.R., May,G., and Bremer,E. (1988) *Cell*, **52**, 569.
53. Freemont,P.S., Lane,A.N., and Sanderson,M.R. (1991) *Biochem. J.*, **278**, 1.
54. Johnson,P.F. and McKnight,S.L. (1989) *Annu. Rev. Biochem.*, **58**, 799.
55. Martin,K., Huo,L. and Schleif,R.F. (1986) *Proc. Natl. Acad. Sci. USA*, **83**, 3654.
56. Dunn,T.M., Hahn,S., Ogden,S., and Schleif,R.F. (1984) *Proc. Natl. Acad. Sci. USA*, **81**, 5017.
57. Sauer,R.T., Jordan,S.R., and Pabo,C.O. (1990) *Adv. Prot. Chem.*, **40**, 1.
58. Hochschild,A. and Ptashne,M. (1986) *Cell*, **44**, 681.
59. Kramer,H., Niemoller,M., Amouyal,M., Revet,B., von Wilcken-Bergmann,B., and Müller-Hill,B. (1987) *EMBO J.*, **6**, 1481.
60. Koudelka,G.B. and Carlson,P. (1992) *Nature*, **355**, 89.
61. Ansari,A.Z., Chael,M.L., and O'Halloran,T.V. (1992) *Nature*, **355**, 87.
62. Frederick,C.A., Grable,J., Melia,M., Samudzi,C., Jen-Jacobson,L., Wang,B.-C., Greene,P., Boyer,H.W., and Rosenberg,J.M. (1984) *Nature*, **309**, 327.
63. Cairns,J. (1963) *J. Mol. Biol.*, **6**, 208.
64. Fuller,R.S. and Kornberg,A. (1983) *Proc. Natl. Acad. Sci. USA*, **80**, 5817.
65. Bramhill,D. and Kornberg,A. (1988) *Cell*, **52**, 743.
66. Schnos,M., Zahn,K., Inman,R.B., and Blattner,F.R. (1988) *Cell*, **52**, 385.
67. Umek,R.M. and Kowalski,D. (1988) *Cell*, **52**, 559.
68. Kim,R.A. and Wang,J.C. (1989) *J. Mol. Biol.*, **208**, 257.
69. Wasserman,S.A. and Cozzarelli,N.R. (1986) *Science*, **232**, 951.
70. Steck,T.R. and Drlica,K. (1984) *Cell*, **36**, 1081.
71. DiNardo,S., Voelkel,K., and Sternglanz,R. (1984) *Proc. Natl. Acad. Sci. USA*, **81**, 2616.

72. Yanagida,M. and Wang,J.C. (1987) *Nucleic Acids Mol. Biol.*, **1**, 196.
73. Minden,J.S. and Marians,K.J. (1985) *J. Biol. Chem.*, **261**, 11906.
74. Landy,A. (1989) *Annu. Rev. Biochem.*, **58**, 913.
75. Better,M., Lu,C., Williams,R.C., and Echols,H. (1982) *Proc. Natl. Acad. Sci. USA*, **79**, 5837.
76. Mizuuchi,K., Fisher,L.M., O'Dea,M.H., and Gellert,M. (1980) *Proc. Natl. Acad. Sci. USA*, **77**, 1847.
77. Mizuuchi,K., Gellert,M., Weisberg,R.A., and Nash,H.A. (1980) *J. Mol. Biol.*, **141**, 485.
78. Pollock,T.J. and Nash,H.A. (1983) *J. Mol. Biol.*, **170**, 1.
79. Spengler,S.J., Stasiak, A., and Cozzarelli,N.R. (1985) *Cell*, **42**, 325.
80. Grindley,N.D.F. and Reed,R.R. (1985) *Annu. Rev. Biochem.*, **54**, 863.
81. Reed,R.R. (1981) *Cell*, **25**, 713.
82. Reed,R.R. and Grindley,N.D.F. (1981) *Cell*, **25**, 721.
83. Krasnow,M.A., Stasiak,A., Spengler,S.J., Dean,F., Koller,T., and Cozzarelli,N.R. (1983) *Nature*, **304**, 559.
84. Wasserman,S.A., Dungan,J.M., and Cozzarelli,N.R. (1985) *Science*, **229**, 171.
85. Benjamin,H.W. and Cozzarelli,N.R. (1988) *EMBO J.*, **7**, 1897.
86. Boocock,M.R., Brown,J.L., and Sherratt,D.J. (1985) *Biochem. Soc. Trans.*, **14**, 214.
87. Craigie,R. and Mizuuchi, K. (1986) *Cell*, **45**, 793.
88. Ptashne,M. (1986) *Nature*, **322**, 697.

Glossary

A-DNA: a conformation of right-handed double-stranded DNA characterized by a tilting of the bases with respect to the helix axis; found in DNA at low humidity.

B-DNA: a conformation of right-handed double-stranded DNA characterized by a perpendicular arrangement of the bases with respect to the helix axis; thought to be the form of DNA commonly found in nature.

Catenane: interlinked double-stranded DNA circles, as in the links of a chain.

Catenane number (Ca): the sum of the intermolecular double-stranded nodes of the DNA molecules comprising a catenane.

Closed-circular DNA: double-helical DNA in which both strands are closed-circles, i.e. there are no free ends.

Cruciform: a stem–loop structure in double-stranded DNA formed by intra-strand base-pairing of each strand.

Curved DNA: DNA in which the preferred conformation has a curved axis.

Flexible DNA: DNA which can be deformed, particularly by bending, more easily than average.

H-DNA: a triple-stranded conformation of DNA comprising a polypyrimidine strand lying in the major groove of a polypurine–polypyrimidine duplex.

Helical repeat (h): the number of base pairs per turn of the double-stranded DNA helix. This can be measured relative to the DNA helical axis (twist-related h) or relative to a surface on which the DNA lies (surface-related h).

Holliday junction: an intermediate in recombination formed by strand-exchange between two homologous duplexes. It is structurally related to a cruciform.

Knot: a knotted length of linear DNA closed into a circle.

Knot number (Kn): the sum of the minimum number of nodes in a knotted DNA circle.

Linking difference (ΔLk): the difference between the linking number of a particular topoisomer of closed-circular DNA and the average linking number of relaxed DNA (i.e. $\Delta Lk = Lk - Lk°$). Positive linking difference ($+\Delta Lk$) corresponds to positively supercoiled DNA and $-\Delta Lk$ to negatively supercoiled.

Linking number (Lk): the number of times the two strands of closed-circular DNA are linked. It is strictly defined as half the sum of the inter-strand nodes in a 2-D projection of the DNA. Linking number is distributed between the two geometric parameters twist (Tw) and writhe (Wr), i.e. $Lk = Tw + Wr$.

Negatively supercoiled DNA: supercoiled DNA with a negative average linking difference.

Node: a crossover point of two DNA single- or double-strands in a 2-D projection. The sign of a node is determined by the directions assigned to the DNA strands.

Open-circular DNA: a double-strand DNA circle containing a broken phosphodiester bond in one strand.

Positively supercoiled DNA: supercoiled DNA with a positive average linking difference.

Recombination: the rearrangement of a DNA molecule or molecules resulting from the breakage of the DNA duplex and its rejoining to another site.

Relaxed DNA: closed-circular DNA formed without any constraint of the DNA helix; it normally consists of a distribution of topoisomers.

Specific linking difference (σ): the linking difference normalized to the length of the DNA, by dividing by $Lk°$ (i.e. $\sigma = \Delta Lk/Lk°$).

Supercoiled DNA: closed-circular DNA formed under a torsional stress. A vernacular term for DNA with a non-zero linking difference. Supercoiling can be manifested as a change in twist and/or a change in writhe.

Surface linking number (SLk): for a closed-circular DNA lying on a surface, SLk is the linking number of the helix axis and a line traced out by the surface normal.

Surface twist (STw): the twist of a line traced out by the surface normal and the DNA axis, for DNA lying on a surface.

Topoisomer: a closed-circular DNA molecule of unique linking number. Molecules differing only in linking number are topoisomers.

Topoisomerase: an enzyme which catalyses changes in the linking number of closed-circular DNA.

Twist (Tw): simply stated, twist is the number of double-helical turns in a given length of DNA, measured relative to the DNA helix axis. In fact, twist cannot normally be determined by counting and is a more complex function of the path of a DNA strand about the helix axis.

Type I topoisomerase: a topoisomerase whose reaction is characterized by linking number changes of ± 1.

Type II topoisomerase: a topoisomerase whose reaction is characterized by linking number changes of ± 2.

Winding number (Φ): the number of double-helical turns in a closed-circular DNA measured relative to a surface on which the DNA helix axis lies.

Writhe (Wr): a parameter describing the path of a DNA helix axis in space. Planar DNA, and DNA whose axis lies on a sphere without crossing, have $Wr = 0$.

Z-DNA: a conformation of double-stranded DNA characterized by a left-handed helix. It can be formed in alternating purine–pyrimidine sequences.

Index